THE HYDROGEN ECONOMY

FUNDAMENTALS,
TECHNOLOGY, ECONOMICS

THE HYDROGEN ECONOMY

FUNDAMENTALS, TECHNOLOGY, ECONOMICS

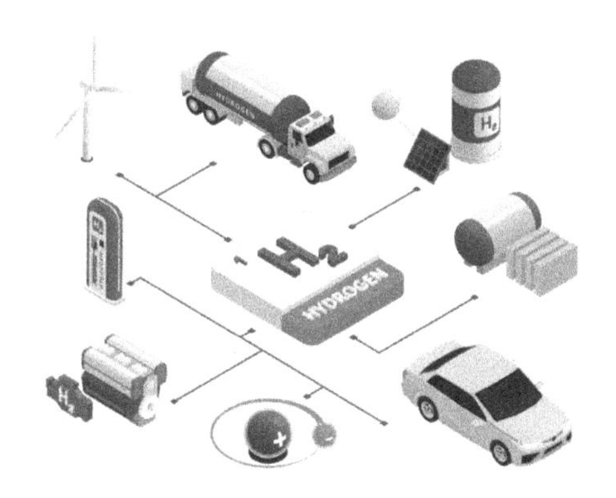

DUNCAN SEDDON
Duncan Seddon & Associates Pty. Ltd, Australia

World Scientific

NEW JERSEY • LONDON • SINGAPORE • BEIJING • SHANGHAI • HONG KONG • TAIPEI • CHENNAI • TOKYO

Published by

World Scientific Publishing Co. Pte. Ltd.
5 Toh Tuck Link, Singapore 596224
USA office: 27 Warren Street, Suite 401-402, Hackensack, NJ 07601
UK office: 57 Shelton Street, Covent Garden, London WC2H 9HE

British Library Cataloguing-in-Publication Data
A catalogue record for this book is available from the British Library.

THE HYDROGEN ECONOMY
Fundamentals, Technology, Economics

ISBN 978-981-124-854-2 (hardcover)
ISBN 978-981-124-855-9 (ebook for institutions)
ISBN 978-981-124-856-6 (ebook for individuals)

For any available supplementary material, please visit
https://www.worldscientific.com/worldscibooks/10.1142/12593#t=suppl

Typeset by Stallion Press
Email: enquiries@stallionpress.com

for Sarah, Laura and Alex

DISCLAIMER

This book is for educational purposes and gives overviews of available technologies and economics for the production of hydrogen in the processing industries. By their nature, these overviews approximate the technology and associated costs. All opinions concerning equipment manufacturers, technology, technology licensors and company strategies or services are the personal opinion of the author and may not necessarily reflect the opinions, positions or claims made by manufacturers or process licensors mentioned herein. All results are based on information available to the author at the time of writing. Changes in factors upon which the overviews are based could affect the results. Forecasts are inherently uncertain because of events and combinations of events that cannot reasonably be foreseen including the actions of governments, individuals, third parties and competitors. No implied warranty or merchantability or fitness for a particular purpose can be given or implied by the author or publisher for any commercial decision based on the contents of this book. Any commercial decision has to be made as a consequence of considerable further work. The information, data and opinions expressed in this book may be affected as a consequence of information, not in the possession of the author.

INTRODUCTION, PREAMBLE, EXECUTIVE SUMMARY

I have spent over 40 years in the chemical process industry and at the prompting of a colleague, I developed a specialist workshop on the 'Hydrogen Economy'. This book is the result of this workshop.

The subject of the 'Hydrogen Economy' is very broad ranging from the potential use of hydrogen for domestic use to the mass production of hydrogen replacing natural gas (LNG) and conventional transport fuels. The 'Hydrogen Economy' or rather the talk about its inevitability seems to be devoid of a rational discussion of methodology and in particular the cost and benefits of its implementation, and the ultimate cost of hydrogen as a fuel replacing carbon emissions from fossil fuels. The subject matter is moving very rapidly and expanding daily and it is impossible to give a comprehensive view. For what is true today may not be true tomorrow. But some aspects of the Hydrogen Economy will always be restrained, particularly by the second law of thermodynamics, which is not always appreciated by the promoters. Just because we can model a perpetual motion machine on a computer does not mean we can make one.

This work focuses on the discussions of the underlying technology and the process economics as they relate today. It attempts to estimate the underlying costs of production and identify the key issues that need improvement to assist the commercial implementation of a Hydrogen Economy. It is important to note that this work does not set out to deliver a 'road map' for commercialisation, rather its role is to ask questions of the current approaches and to inform those concerned with its implementation: it sets out to identify hurdles to implementation that may not be apparent to those entering the field for the first time.

For the hydrogen economy, we need hydrogen. On Earth, the only two major sources of hydrogen are water and hydrocarbons; its concentration in the atmosphere is negligible and hydrogen in natural gas is rare. Water is the most common source as water covers 70% of the Earth's surface and by rain it falls to earth in a relatively pure form. The hydrogen content of water is 11.1% by weight but extracting it is the big problem. The hydrogen–oxygen bond of water is very strong at 484 kJ/mol. This bond can be broken by electrolysis, liberating hydrogen and oxygen, or by reaction with hydrocarbons such as methane.

Methane itself is a source of hydrogen, a carbon atom is bound by four strong carbon–hydrogen bonds at 438 kJ/mol. However, unlike water, which requires plasma range temperatures of several thousand degrees to decompose into the elements, methane can be decomposed to carbon and hydrogen at more modest temperatures (1000 K). However, a better and commercialised approach is by reacting the methane with water in a process known as steam-methane-reforming (SMR), which delivers hydrogen and carbon oxides. If the carbon oxides are captured and stored, the resultant hydrogen is often referred to as 'blue' hydrogen, whilst if the hydrogen is produced from electrolysis using electricity from renewable sources the hydrogen is referred to as 'green' hydrogen.

The book discusses the cost of production from these two sources and seeks to explain the fundamental technology and the underlying process economics.

The author is indebted to Judith for assistance in editing the manuscript and to Dr Michael Clarke for informative discussions and helpful suggestions.

SUMMARY OBSERVATIONS

Production

Blue hydrogen

- The lowest cost approach for hydrogen generation avoiding carbon emissions to the atmosphere is blue hydrogen.
- Technology selection can be used to optimise carbon capture and storage, which is largely proven commercially via carbon dioxide use in enhanced oil recovery. The main problem is the cost. Cost remains a significant issue for extracting carbon dioxide from flue gas.
- For optimum economic performance, the manufacturing facility should be located near a source of low-cost natural gas and depleted gas fields and saline aquifers could also be used for carbon disposal.

Green hydrogen

- Despite its simplicity, electrolysis is not very efficient. There are major scientific and technical challenges to improving the underlying efficiency, particularly to the fundamental chemistry of how the oxygen is liberated in the cell.
- With present technology, to be competitive green hydrogen requires very low power prices, typical of that delivered by large-scale hydrogeneration at about \$20/MWh.

- The economics would be significantly improved if the efficiency of water electrolysis were improved. This would lower both capital and operating costs and water efficiency.

Storage and transport

- The lowest cost storage option is in salt caverns. Salt beds are common in oil and gas provinces and could be developed for hydrogen storage in blue hydrogen projects.
- Production of liquid hydrogen is costly.
- It is not clear if present LNG technology and shipping logistics could be adapted for liquid hydrogen.
- In theory, mass transport of liquid hydrogen by ship is a low-cost means of transport — the main issue is the high cost of liquefaction.
- Transport of hydrogen as an intermediate such as naphthene or ammonia is feasible and has been demonstrated.
- The significant cost of using intermediates lies in the loss of the intermediate when regenerating the hydrogen. This requires costly make-up (naphthenes) or loss due to adverse heating requirements (ammonia).
- Widespread use of ammonia as a transport medium is likely to bring with it widespread safety issues.
- Long distance dedicated hydrogen pipelines are commercially proven. Where it is feasible, pipelines offer a cost-competitive manner for transporting hydrogen over long distances.
- There is considerable interest in using natural gas pipelines for the carriage of hydrogen. This may be an option in special situations. This option is cheaper than dedicated hydrogen pipelines but is likely to lead to higher losses and downstream users of natural gas may be compromised if the hydrogen is not removed prior to delivery.
- Storage and transport of hydrogen as compressed gas is only viable for niche operations or short transport distances.

Competitive position

Stationary applications

- With the present structure of the energy market, hydrogen, apart from special considerations, is uncompetitive against currently used fuels for large stationary power generation.
- Hydrogen will remain uncompetitive without high carbon taxes, that is carbon emission charges of more than $100/t$CO_2$, or government directives or a combination. This will have a major adverse impact on power prices to the consumer in transitioning from a fossil fuel.
- This is also the case for direct substitution in conventional internal combustion engines.

Vehicles

- The Hydrogen Economy proposes to use hydrogen in a fuel cell to displace internal combustion engines. Fuel cells have higher efficiency than internal combustion engines. This advantage makes them feasible provided hydrogen can be delivered at the low cost expected for an optimum blue hydrogen scenario.
- Because volumetric energy density is a key factor, on a volume for volume basis hydrogen fuel will not deliver the mileage delivered by the current optimum gasoline or diesel engines. However, hydrogen will deliver better mileage than current battery technology.
- For vehicle fuel use the favoured method of on-board storage is compressed gas to a minimum of 35 MPa, preferably 70 MPa (700 bar). Compression to these levels is expensive and energy intensive and will demand significant quantities of power.
- Furthermore, although widespread in the present process industries, the engineering to this high hydrogen pressure for general use would require a significant level of training and safety awareness for the engineers and technicians concerned with its implementation and facility servicing.

Dr. Duncan Seddon, FRACI, CChem.
July 2021

CONTENTS

The Dakota Coal Gasification Company, Great Plains Synfuels Facility at Beulah, ND, USA produces about 150 MMscfd of methane (equivalent to 450 MMscfd hydrogen) from lignite. Carbon dioxide (2.3–3 Mt/y) is sent by a 205-mile pipeline to a carbon capture and storage facility in Canada. Photograph care of the Dakota Coal Gasification Company.

World scale ammonia facility producing 1200 t/d ammonia at Misr Oil Processing Company, Damietta, Egypt. The picture illustrates the large methane steam reformer box and in the foreground the secondary reformer and the converter. ThyssenKrupp is thanked for the photograph; ©thyssenkrupp.

Large scale membrane electrolysis cell operation at Tessenderlo, Belgium. The facility has a capacity of 306,000 t/y sodium hydroxide and 272,000 t/y chlorine. World-scale water electrolysis for the hydrogen economy would be much larger. ThyssenKrupp is thanked for the photograph; ©thyssenkrupp

World scale coal gasification plant at Puertollano, Spain. The facility is for IGCC power generation and produces 157,000 m^3/h of CO + H_2 using an entrained bed gasifier (Prenflo). ThyssenKrupp is thanked for the photograph; ©thyssenkrupp

A large-scale natural gas-based DRI facility at Gubkin, Belgorod Oblast, Russia. In the foreground is a Midrex LGOK HBI-3 plant clearly showing the tall shaft furnace. Behind and to the right are HBI-2 and HBI-1. The pelleting/beneficiation plant is further back to the right. The ore mine is to the left in the background. Midrex corporation claims that these operations can be built or adapted to use hydrogen as a reducing gas. Midrex Corporation is kindly thanked for the photograph; ©midrexcorporation.

CHAPTER 1

THE MARKET FOR HYDROGEN

The Use of Hydrogen in the Process Industries

The use of hydrogen in the process industries is spread across the refining, petrochemical, chemical and metallurgical industries. It is dominantly produced on-site for a particular process at the purity required by the process. Often, it is made as a combination with carbon monoxide; the hydrogen carbon monoxide mixture being known commonly as synthesis-gas. In some instances, where demand is large across a geographic region, such as the US Gulf, it is produced by third-party players (the merchant gas companies) who deliver the product by pipeline to several users. For smaller users, these companies (the merchant gas companies) provide hydrogen in compressed tanks delivered by trucks. In this chapter, we are mainly concerned with the mass production and use of hydrogen in the process industries, the many smaller uses for hydrogen for specific uses are not discussed.

Safety

The hydrocarbon process and chemical industries are classified as hazardous operations, the staff and operators are trained to a high level of competence and safety awareness. This includes the necessary training for the production and use of hydrogen often in large amounts and at high pressure.

Hydrogen and other chemicals, which may find a role in the Hydrogen Economy, such as ammonia are highly hazardous materials and require specialized training and expertise. In the process industries, there have been many cases of fires, explosions and deaths resulting from accidents using these materials.[1] The role of extensive safety awareness and training cannot be overestimated in all aspects of a Hydrogen Economy.

Oil Refineries

The purpose of a crude oil refinery is to convert crude oil into transport fuels, particularly gasoline, jet fuel and motor diesel, which have specific properties that are generally specified in quality by government regulations. It is useful to consider the role of refineries by noting the hydrogen to carbon molar ratio of crude oil and the principal products, as is illustrated in Figure 1.1.

Crude oil has a typical hydrogen to carbon molar ratio (H/C) in the range of 1.4–1.6. Heavy crude oils and synthetic oils (oils derived from bitumen) can be below this range. In the extreme, coal has a H/C ratio of about one and has in the past been used to produce transport fuels but is not relevant to this discussion.[2]

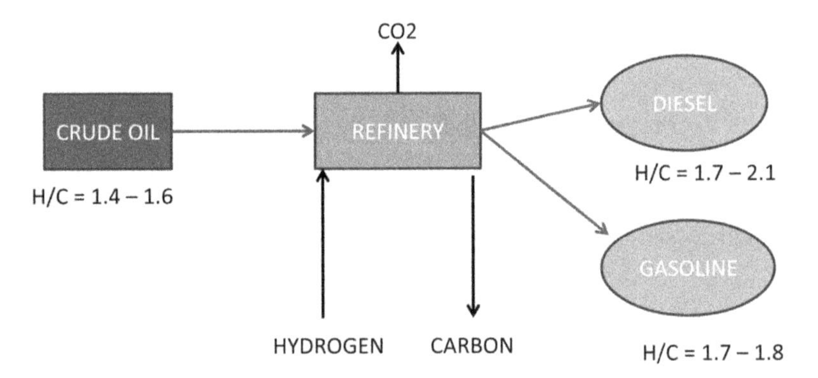

Fig. 1.1: The hydrogen/carbon ratio of crude oil and diesel and gasoline.

[1] Kletz TA. (1988) *What Went Wrong — Case Histories of Process Plant Disasters*, 2nd ed. Gulf Publishing, Houston.

[2] For example, by the Fisher-Tropsch process where coal is converted into synthesis gas and then on to fuels.

The transport fuels diesel (and jet-fuel) made to current specifications have a hydrogen to carbon molar ratio (H/C) nearer to 2 and gasoline is in the range 1.7–1.8. These ranges have changed over the years to increase the hydrogen content. Most jurisdictions have demanded fuels of higher hydrogen content for environmental reasons, for example by lowering the amount of sulphur and the quantity of aromatic molecules that are present in the final specification fuel.

The tightening of fuel specifications along these lines is likely to continue, thus requiring more hydrogen to be used (or, which is less likely, more carbon to be rejected). Furthermore, most remaining crude oil reserves are at the heavier end of the crude oil spectrum that requires more hydrogen to process.

In converting a feed with a H/C of 1.6 or less to fuels with a H/C nearer to two, the refinery can add hydrogen, eliminate carbon as carbon itself or as carbon dioxide. The route chosen is dependent on local factors, many large refinery operations use all three routes. All refineries are different using different combinations of unit processes, depending on the location, the local market for fuels and the crude oils being processed. With this in mind, a typical modern refinery operation is illustrated in Figure 1.2.[3]

Crude oil enters the refinery, and after some preliminary treatment such as water and salt removal (not shown), is heated and passed to a distillation column. The distillation operates near atmospheric pressure where the crude oil is separated into fractions of differing boiling point. The light gaseous fractions are passed to the gas plant that separates propane and butane (fuel gas and LPG), some of which are used to produce alkylate (a gasoline blending component). Butane is used in the final stages of gasoline production to meet vapour pressure requirements.

The liquid fractions are separated broadly into four groups:

- boiling point 30–196°C for blending into gasoline, generally referred to as naphtha,
- boiling point 196–235°C for blending into kerosene or jet-fuel (not shown in the figure),

[3]There are several handbooks that describe in some detail refineries, refinery unit operation, and refinery economics. I have used data provided in Gary JH, Handwerk GE. (2001) *Petroleum Refining — Technology and Economics*, 4th ed. Marcel Dekker Inc., New York and Maples RE. (1993) *Petroleum Refinery Process Economics*. PennWell, Tulsa, OK.

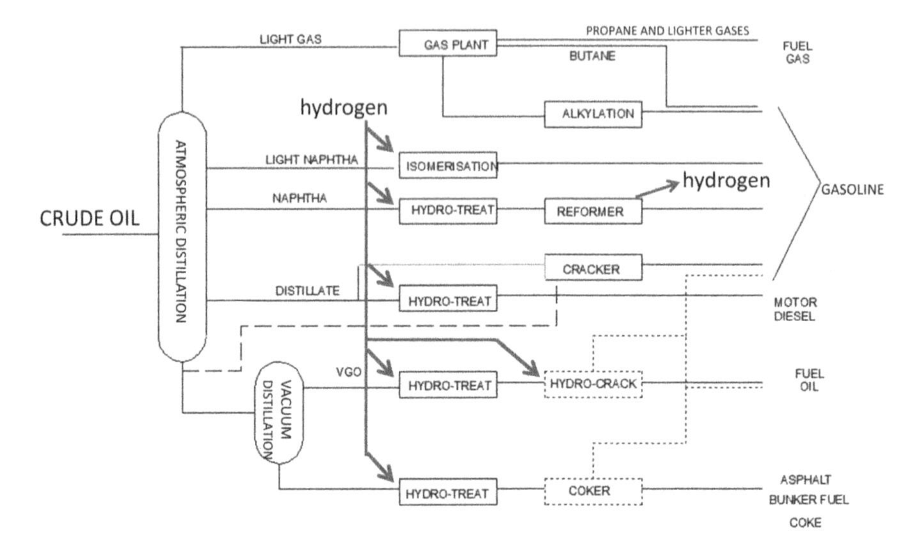

Fig. 1.2: Refinery operations requiring hydrogen.

- boiling point 235–360°C for blending into motor diesel,
- boiling point >360°C for fuel oil or further processing.

The fuel oil fraction is often passed to a vacuum distillation column which separates fractions (VGO, vacuum gas oil) with boiling range up to about 550°C.

All these primary fractions require hydro-treatment to improve the efficacy of downstream processes and make them suitable for fuel blending by removing sulphur and other elements. This hydrogen requirement is indicated by the downward facing arrows in the figure. The more important unit operations are discussed below.

Naphtha Isomerization

A typical process layout is shown in Figure 1.3.

The duty of the naphtha isomerisation unit is to increase the octane number of the light naphtha fraction coming from the atmospheric distillation column. After drying, the naphtha is heated (to typically 200°C) and passed to a reactor (typically charged with a platinum supported on

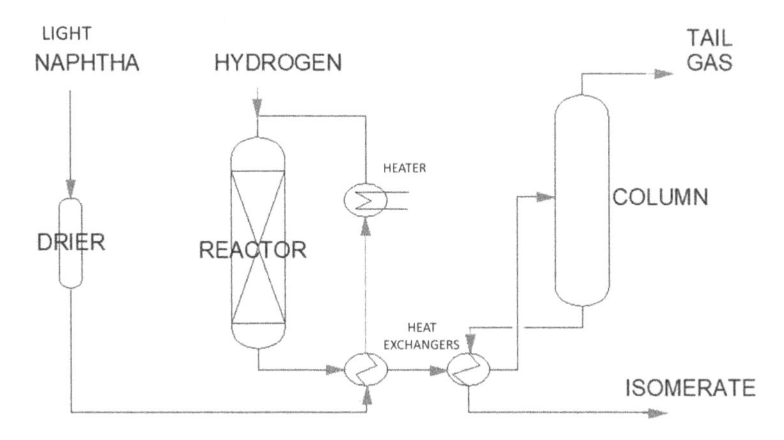

Fig. 1.3: Naphtha isomerisation.

alumina catalyst) which brings about the required isomerisation of linear (low octane) hydrocarbons to branched (high octane) hydrocarbons. To improve the stability of the catalysts and prolong their life, the catalyst best works in the presence of hydrogen which is added to the system. After distillation to separate the higher octane isomerate, the hydrogen in the tail gas can be separated and recycled.

Hydrogen consumption is low, typically 40scf/bbl of fresh feed.

Naphtha and Distillate Hydro-Treating

The aim of hydro-treatment is to remove elements, particularly sulphur, from refinery streams prior to use as blendstock or to improve the operation of downstream processes. It is also often referred to as HDS (hydrodesulphurisation). The use of hydro-treatment to meet the tighter fuel specifications for gasoline and diesel (less than 10 ppm sulphur) has increased dramatically in the past decade. There are other benefits of the process:

- removal of other elements — nitrogen and oxygen — from the fuel,
- removal of unsaturated molecules (olefins and diolefins) — this improves fuel stability (prevents gum formation) and resistance to oxidation. This is especially important for petrochemical feedstock,

- reduce or remove aromatics to meet tighter specifications on benzene or poly nuclear aromatics (PNAs) and to improve diesel cetane number.

Hydro-treatment also has the advantage of lowering fuel density i.e. increasing fuel volume per unit mass which has advantages as the sales of refinery products are universally sold on a volumetric basis rather than a mass basis.

A typical hydro-treatment layout is shown in Figure 1.4, which illustrates the process for gas oil fractions (b.p. 235–360°C). Some processes, such as thermal cracking and coker operations, produce streams laden with diolefins which require treatment in a separate hydrogenation reactor. The gas oil streams are heated to about 320–400°C and hydrogen at 2–6 MPa pressure is added. The main reactor is charged with a catalyst-based cobalt-molybdenum supported on alumina (Co/Mo/Al_2O_3; there are many variants). Hydrogen sulphide produced in the process is washed out in the separation stages prior to distillation.

There are many variations of hydro-treatment ranging from naphtha hydro-treatment for preparing heavy naphtha fractions (b.p. 100–196°C) for naphtha reforming to heavy gas-oil hydro-treatment for diesel blending.

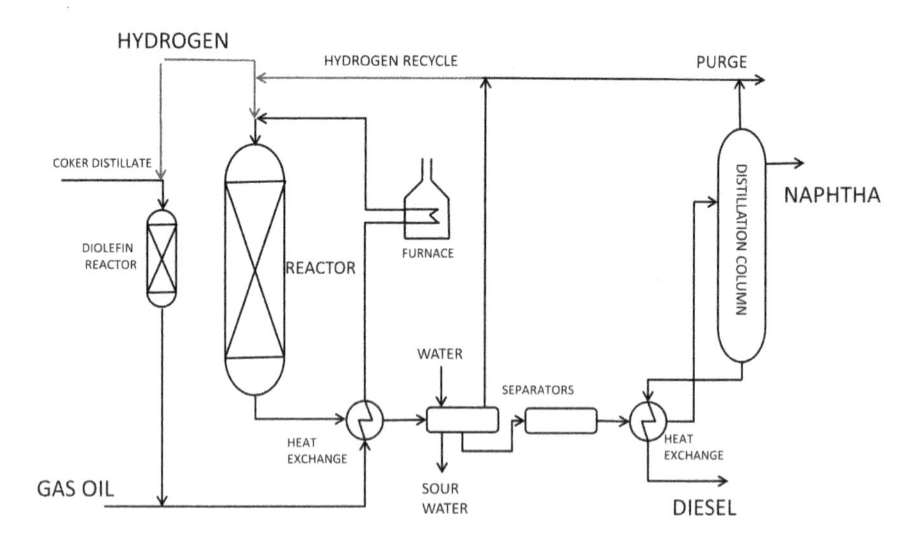

Fig. 1.4: Gas oil hydro-treater.

Some hydro-treating operations also have the duty to reduce aromatics. This generally requires higher pressure operation (to 10 MPa).

Hydrogen consumption varies as to feedstock and duty. Table 1.1 gives typical values.

Table 1.1: Typical hydrogen requirement for hydro-treatment

Feed	Heavy distillate	Light distillate	Naphtha
Hydrogen (scf/bbl)[4]	400–800	150–400	100–150

Hydro-Cracking

The hydrocracker has the duty to take heavy (high boiling) gas oil and fuel streams from the atmospheric distillation column or the vacuum distillation column and convert these into lighter streams for fuel blending.

A typical layout is given in Figure 1.5, which shows the process for a heavy gas oil. The process is conducted in two stages. The first stage is very similar to hydro-treatment discussed above and involves pre-treatment of any coker or similar streams containing diolefins. The first stage uses a Co/Mo/Al_2O_3 catalyst with the duty to remove sulphur and

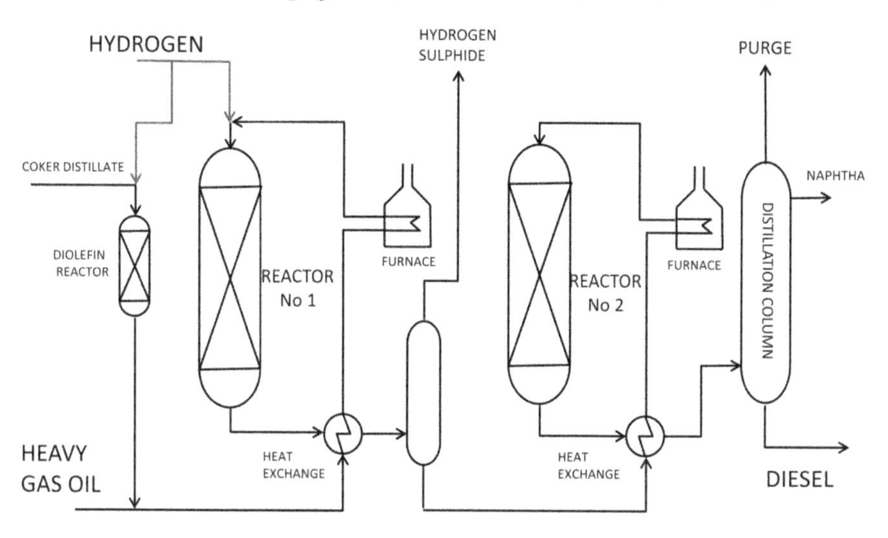

Fig. 1.5: Layout for two-stage hydrocracker.

[4] Hydrogen consumption in standard cubic feet per barrel of feed (scf/bbl).

other materials for the efficient operation of the second stage which uses typically a platinum catalyst supported on an acidic zeolite catalyst which cracks (breaks) the larger molecules in the gas-oil into smaller molecules boiling in the naphtha and diesel boiling ranges; a particular emphasis is on the production of the diesel fraction.

For hydro-cracking, the operating temperature is somewhat higher than that of a hydro-treater (400–500°C) and the pressure is considerably higher for most feedstocks (20 MPa). Judicious selection of feedstock can reduce the required operating pressure to about 10 MPa.

Hydro-crackers can be operated at various levels of severity depending on the requirements of the refinery. Typical hydrogen consumption is in the range 1000–3000 scf/bbl of feed.

The Naphtha Reformer

The naphtha reformer has the duty to covert low octane heavy naphtha into high octane reformate for gasoline blending. In this process, aliphatic molecules (paraffins) and alicyclic molecules (naphthenes) are converted into aromatics — benzene, toluene and xylene. In this process, hydrogen is released and is the main manufacturing process for hydrogen used in the hydrogen-consuming units within the refinery. This is indicated by the upward-facing arrow in Figure 1.2.

The chemistry of the process is illustrated below[5] for *n*-heptane which has a Research Octane Number (RON) of 0. This is progressively converted via a naphthene (RON 86) to toluene (RON 124).

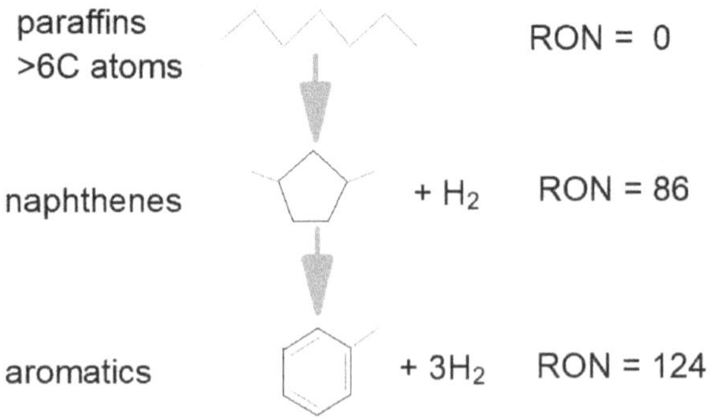

paraffins >6C atoms — RON = 0

naphthenes — $+ H_2$ — RON = 86

aromatics — $+ 3H_2$ — RON = 124

---

[5]The illustration uses the standard stick notation for organic molecules. Only carbon–carbon bonds and hetro-atoms are shown. All other valence bonds are carbon–hydrogen.

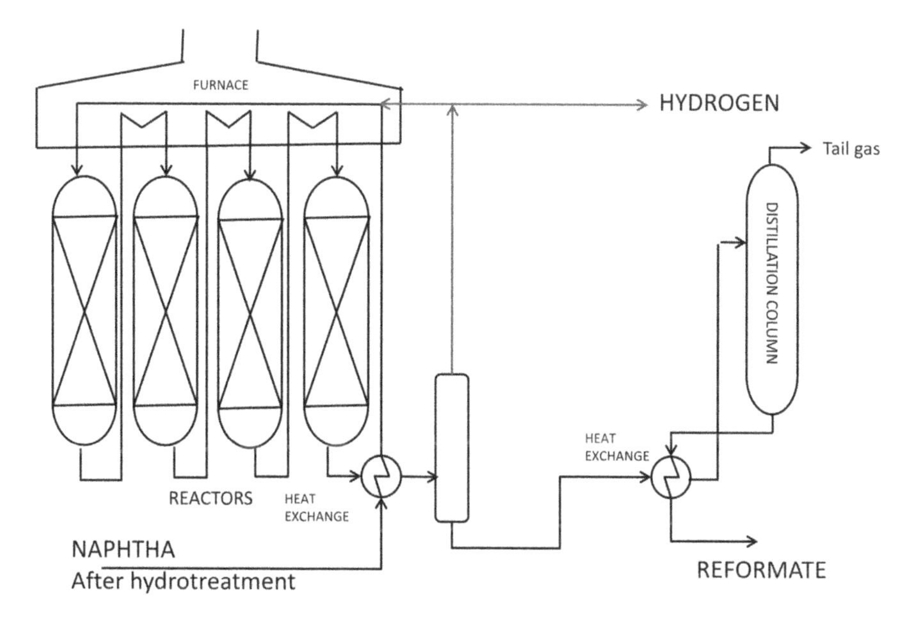

Fig. 1.6: Naphtha reforming (semi-regenerative technology).

There are several variants on the technology but a common process is illustrated in Figure 1.6.

First, naphtha is hydro-treated to remove sulphur and other impurities that would poison or inhibit the sensitive catalyst. This is typically based on platinum supported on alumina and promoted with rhenium. The treated naphtha is mixed with hydrogen and heated in a furnace to about 500°C and passed to a reactor. The reaction is highly endothermic which cools the fluid streams which have to be reheated to the reaction temperature. This process is repeated through several reactors. After the final reactor, the fluids are cooled and the excess hydrogen is separated prior to separating the reformate from light hydrocarbon gases produced in the process.

As the process proceeds, the catalyst deactivates and the reactors or catalyst is taken off-line for regeneration. There are several proprietary alternatives for the process.

Hydrogen production is a function of the severity of the operation — the quantum of aromatics produced in the process measured by the octane number of the reformate. Typical hydrogen yields range from about 500 scf/bbl to over 1300 scf/bbl of feed for a reformate of 100 RON (Research Octane Number).

The reforming process could find a useful role in facilitating the transport of hydrogen across the seas; other aspects of the process are discussed in further chapters.

The Refinery Hydrogen Market

The *Oil & Gas Journal* compiles a list of world refinery operations. The unit operations and their capacities are included; the list is published annually. Table 1.2 gives an analysis of the 2008 and 2020 data.

The table lists the units requiring hydrogen and the range of the demand and a typical value. This latter value is used with the quoted average production rates to estimate the hydrogen demand for the years 2008

Table 1.2: Estimated hydrogen use in refineries from unit capacity data.

Refinery operations			**2008**	**2008 est.**	**2020**	**2020 est.**
Hydrogen consumption	MMscfd/bbl	Average value	MMbbl/d	MMscfd H_2[a]	MMbbl/d	MMscfd H_2
Naphtha isomerisation	40	40	1.756	70	1.841	74
Naphtha hydro-treating	100–150	125	n.a.[b]	n.a.[b]	18.471	2309
Light distillate hydro-treating	150–400	**275**	44.011	12,103	18.116	4982
Heavy distillate hydro-treating	400–800	600	n.a.[b]	n.a.[b]	11.914	7148
Hydro-cracking	1000–3000	2000	4.956	9912	7.508	15,016
Totals			50.723	22,085	57.85	29,529
Reformer H_2 production	500–1300	900	11.431	10,288	11.761	10,585
Hydrogen shortfall				11,797		18,944
Hydrogen plant actual				13,460		16,411
Crude capacity (bbl/d)			85,308		91,714	

(*a*) 1 MMscfd = 28,261 m^3/d; see Appendix for abbreviations; (*b*) n.a.: not available.

and 2020, which is 22,085 MMscfd (624 Mm^3/d) and 29,529 MMscfd (834 Mm^3/d) respectively.[6] Most hydrogen is supplied by naphtha reforming and an estimate for this production is supplied. The hydrogen shortfall is supplied by an in-house hydrogen plant and/or imports. The actual estimated volume of hydrogen production within the refineries is given in the last row and this approximates to the shortfall.[7]

The 29,529 MMscfd hydrogen is approximately 23.6 Mt/y (million tonnes per year) of hydrogen. At a nominal value of $3000/t this makes the value of hydrogen in refinery operations approximately $69 billion.

Petrochemicals

Petrochemical facilities[8] are often juxtaposed to, or close to, refinery operations and are often fully integrated operations. Petrochemicals operations rely on upstream oil and gas operations as well as refinery operations for their feedstock. The principal petrochemical feedstocks are ethane, liquefied petroleum gases (LPG, principally propane and *n*-butane), naphtha, vacuum gas oil and atmospheric residual fuel oil (also known as atmospheric residua); Figure 1.7.

Ethane and LPG are almost exclusively sourced from natural gas. These are primarily used to produce ethylene, which is then converted into poly-olefins (polymer resins — HDPE, LDPE, LLDPE). Both feeds and products have H/C (molar) ratio of slightly more than 2 so there is little call for hydrogen other than for final stabilisation of the products by reducing any excess olefin in the resin.

One of the principal feedstocks is naphtha which is often split into a light and a heavy fraction. Naphtha can be obtained from natural gas condensate as well as cuts from the refinery atmospheric crude oil column. Naphtha is used for the production of olefins and aromatics by steam

[6] Note the data for 2020 is more detailed than for 2008.

[7] Note the data for hydrogen plant capacities reported for 2020 appears in error due to anomalies for data for Austria and Italy. The values for these countries has been removed making the figure for 2020 low.

[8] Seddon D. *Petrochemical Economics — Technology Selection in a Carbon Constrained World*, Imperial College Press, London, (2010).

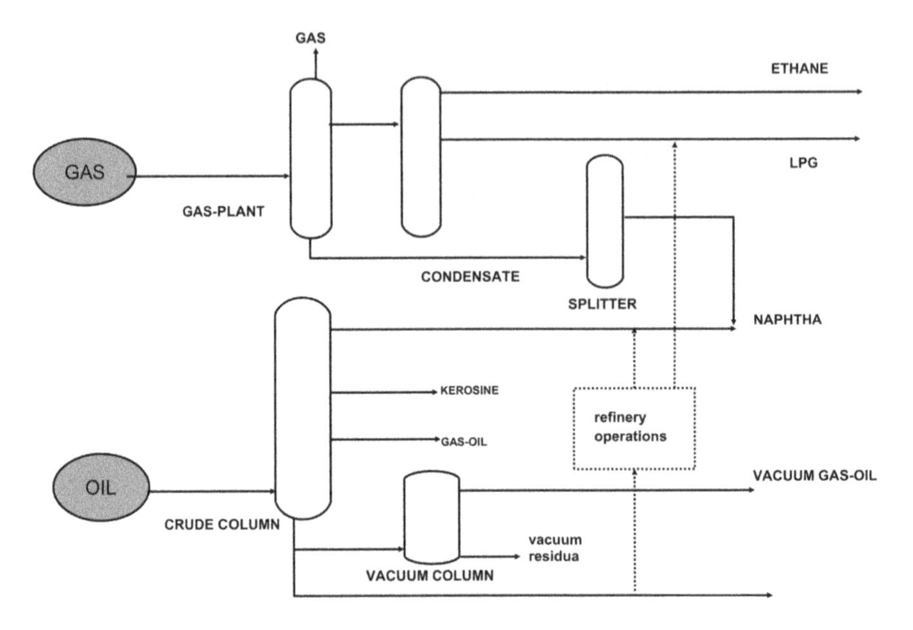

Fig. 1.7: Principal feedstocks for petrochemical operations.

cracking and for the production of aromatics by reforming. Naphtha for petrochemical operations is often selected by specific properties such as the UOP-K factor (Watson-K factor) or the quantum of naphthenes and aromatics present. Light naphtha with a high level of linear paraffin is a good steam cracker feed for light olefins, particularly ethylene and propylene. Heavy naphtha with high naphthene and aromatics content is selected for aromatics production. Steam cracking heavy naphtha produces a wide range of olefins (ethylene, propylene, butenes and butadiene) and mixed aromatics as pyrolysis gasoline.

Vacuum gas oil (VGO) and increasingly atmospheric residual fuel oil (especially derived from waxy, paraffinic crude oil) are used as a steam cracker feedstock to produce olefins and aromatics.

Aromatics, with a very low H/C molar ratio of 1 to 1.1, can only be produced from feedstocks such as naphtha with a significantly higher H/C ratio (1.7 say) by the elimination of hydrogen. This occurs as discussed above in the naphtha reforming process and in the process of steam cracking which also produces significant quantities of by-product hydrogen. These processes generally produce sufficient hydrogen for the petrochemical operational needs without major additional hydrogen production plants.

Naphtha Hydro-Treating

As has been discussed above, it is necessary to hydro-treat naphtha prior to naphtha reforming for the production of aromatics. It is becoming increasingly common to hydro-treat naphtha destined for steam cracking operations. This improves on-stream time in the cracker furnaces. Sometimes natural gas condensate is quoted as a feedstock for steam cracking. This is generally hydro-treated with the final distillation removing the heavier components which would otherwise prove problematic in promoting coking in the cracker furnaces.

Hydrogen Use in Down-Stream Operations

Hydrogen is used in the production of polymers (macromolecules) from light olefins. It is added to the polymerisation system to act as a chain-transfer agent where it reduces the final molecular weight by terminating the polymerisation process and starting the growth of a new chain. It also acts to remove olefin groups that would otherwise occur at the end of the molecular chain. This improves the oxidation stability of the final polymer. However, the use of hydrogen for this purpose is small.

Naphtha Reforming and Aromatics

As discussed in the refining section above, the process of naphtha reforming to convert aliphatic molecules to aromatics produces a significant amount of hydrogen. The duty of the naphtha reformer in the petrochemical context is to produce benzene, toluene and xylene (BTX). This is produced by high severity reforming to maximise the BTX yield that inevitably increases the hydrogen yield. The reformate is passed to an aromatic extraction plant that separates the BTX which is further separated into benzene, toluene and mixed xylenes.

Benzene

Benzene is a major petrochemical intermediate. It is used for the production of styrene that is used for the production of polystyrene and styrene co-polymers such as ABS (acrylonitrile — butadiene — styrene). It is also used in the production of *cyclo*-hexane.

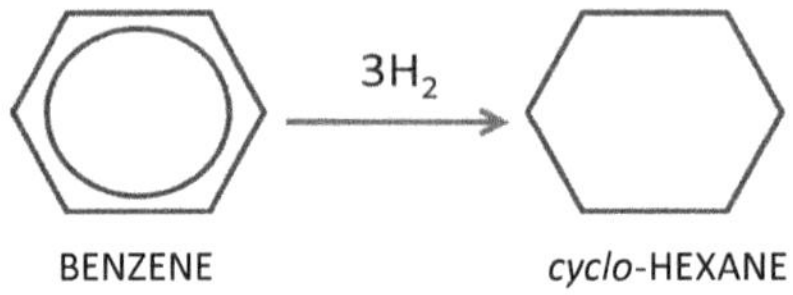

$3H_2$

BENZENE *cyclo*-HEXANE

This process could play a role in the shipping of hydrogen which is discussed in later chapters. *Cyclo*-hexane is a key intermediate in the production of nylon-6 and nylon-6,6.

Toluene

Toluene has few uses in the petrochemicals industry. It is used for the production of xylenes by methylation or disproportionation and for producing benzene and methane by hydro-demethylation.

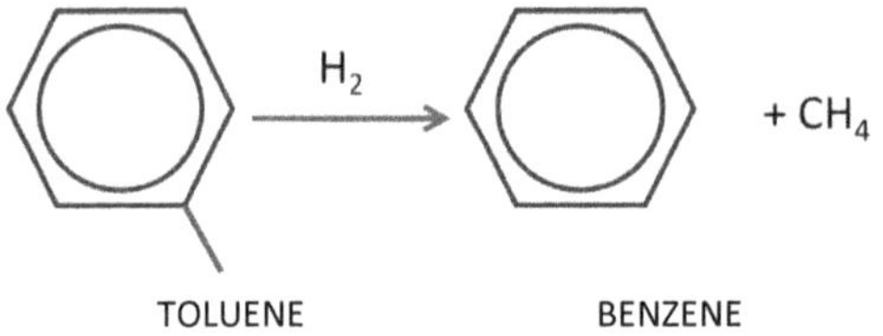

H_2 $+ CH_4$

TOLUENE BENZENE

Xylenes

The term 'xylenes' covers the C_8 aromatics, the three xylene isomers, *meta*-xylene, *ortho*-xylene, *para*-xylene and ethylbenzene. The mixture is often referred to as virgin xylene and is formed in proportion to its thermodynamic stability. Of the xylene isomers, *meta*-xylene is about 50% of the mixture with *ortho*- and *para*- at about 25% each. *Meta*-xylene has little use in the petrochemical industry, and there is extensive plant isomerising the mixture and separating the more useful *ortho*-xylene and in particular *para*-xylene that is in high demand to produce poly-ethylene-terephthalate (PET) for bottles and fiber production.

Ethylbenzene can be used to make styrene with the process generating a molecule of hydrogen.

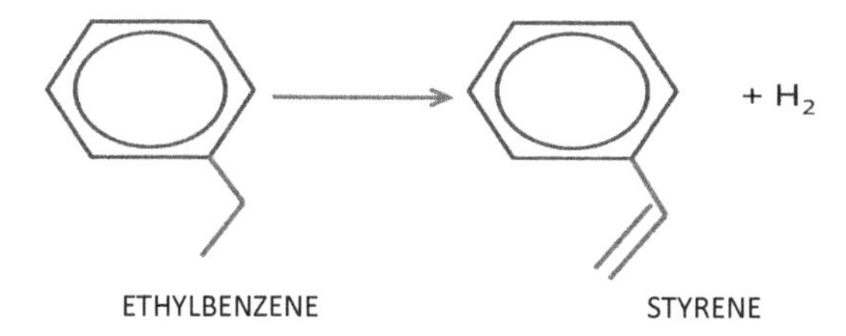

Oxo-alcohols

There are several other uses for hydrogen in the petrochemicals industry: one of which is the production of alcohols by the OXO-process. In this process, an *alpha*-olefin reacts with carbon monoxide and hydrogen to produce an aldehyde, which is then reduced with hydrogen to produce alcohol:

$$\text{R–CH=CH}_2 + \text{CO} \xrightarrow{\text{H}_2/\text{CO}} \text{R–CH}_2\text{–CH}_2\text{–CHO}$$

$$\text{R–CH}_2\text{–CH}_2\text{–CHO} \xrightarrow{\text{H}_2} \text{R–CH}_2\text{–CH}_2\text{–CH}_2\text{–OH}$$

Where R is a linear alkyl group

For OXO alcohols with carbon numbers over 8, the products are further processed to produce detergents. The use of a mixture of carbon monoxide and hydrogen is commonly called synthesis gas and is widely used in the chemicals industry as is discussed in the next section.

The Use of Hydrogen in the Chemicals Industry

Hydrogen is widely used in the heavy (large scale) chemicals industry as a feedstock for the production of methanol and ammonia and their derivatives.[9] In this role hydrogen is for the most part used in combination with

[9] Seddon D. (2006) *Gas Usage and Value — The Technology and Economics of Natural Gas Use in the Process Industries*. PennWell, Tulsa, OK.

carbon monoxide and the two are generally known as synthesis gas. The term synthesis gas covers any combination of hydrogen and carbon monoxide and sometimes includes carbon dioxide. The hydrogen/carbon monoxide molar ratio is referred to as the stoichiometric ratio (SR) of the synthesis gas.

Methanol (CH_3OH)

The world production of methanol is about 60 Mt/y, and it is widely used for the production of formaldehyde (for urea-formaldehyde resins), acetic acid, a range of solvents and as a fuel additive (as methanol itself or as MTBE). The process reaction is as follows:

$$\mathbf{CO + 2H_2 = CH_3OH}$$

There are various proprietary technologies for the production of methanol that are all similar. The process layout is illustrated in Figure 1.8.

In a typical methanol process, synthesis gas (syn-gas) is compressed to about 10 MPa (100 bar) and mixed with recycle gas and heated (not shown) to typically 250°C) prior to passing through a converter charged with

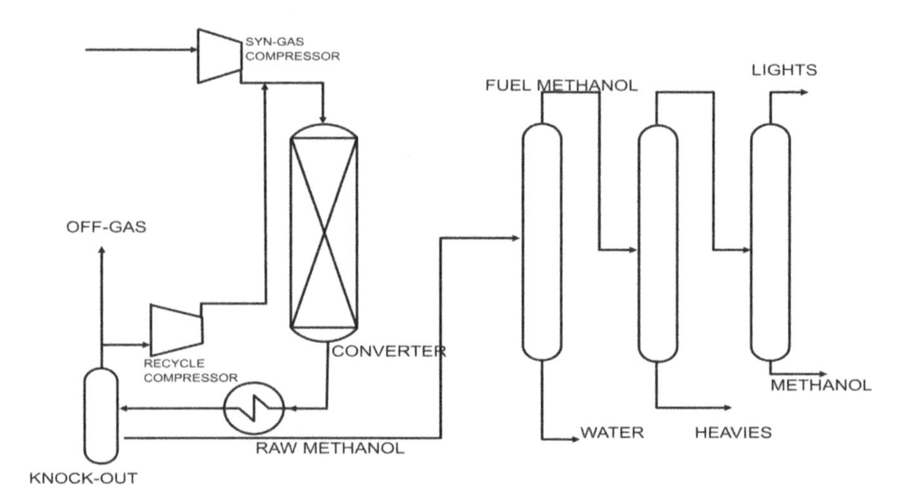

Fig. 1.8: Methanol process.

catalysts ($Cu/ZnO/Al_2O_3$ is the most common) which convert the synthesis gas into methanol. The reaction is very exothermic, and there is a large variety in converter designs to handle the excess heat, which is used to raise steam. The pass conversion is about 15% so there is a considerable recycle stream. After cooling the raw methanol liquid is separated and passed to a primary distillation column to produce fuel-grade methanol. Pure methanol (Federal AA Grade) is produced by further distillation that removes heavier and lighter impurities.

Many older plants use steam reforming of natural gas to produce the synthesis gas. This has a higher SR than that required in the synthesis, the extra hydrogen builds up in the recycle loop and is purged from the system in the off-gas. This hydrogen off-gas is generally used as a fuel in the steam reformer. The older plant may therefore be a source of hydrogen. The excess hydrogen increases the load on the recycle compressor. To lower operating costs, more modern plants have a more compatible SR for the methanol synthesis (SR = 2) that reduces the necessity to purge the excess hydrogen from the synthesis loop.

Methanol and the hydrogen economy

Methanol can be used as a source of pure hydrogen for fuel cells or rocket fuel. The main route is by methanol steam reforming. This produces hydrogen and carbon dioxide. The hydrogen is separated by means of a pressure swing adsorption (PSA), a hydrogen permeable membrane or palladium alloy.[10]

$$\mathbf{CH_3OH + H_2O = CO_2 + 3H_2}$$

The electrolysis of aqueous methanol can be used to produce hydrogen and carbon dioxide as a by-product. This method consumes about half the power of a typical water electrolysis system.[11] This technology is discussed further in a later chapter.

---

[10] Olah GA. (2005) Beyond oil and gas: The methanol economy. *Angew Chem Int Ed*, **11**(18): 2636.

[11] Narayanan S, Chun W, Jefferies-Nakamura B, Valdez TI. (June 2002) *NASA Tech Briefs* NPO-19948.

Ammonia

The world production of ammonia is in excess of 150 Mt/y, and it is mostly used for fertilizer manufacture. For many countries in the world, ammonia production is seen as a strategic chemical and its production and sale are tightly controlled so that only about 20% of the world production is open to 'free trade'.

As fertilizer ammonia is mainly used:

- As ammonia gas or as a water solution. This is fed directly to the roots of growing plants and gives a rapid boost,
- As urea, a solid, which is easy to transport in bags or in bulk road tankers or ships. Because of its easy transport logistics, it is the major source of bulk ammonia production. As a solid or solution urea is relatively easy to apply. In production, soil bacteria breakdown the urea into ammonia and carbon dioxide. The increasing concern to control and eliminate carbon dioxide emissions might mean that the use of urea as a nitrogenous fertilizer is considerably reduced in future years. To maintain agricultural productivity this would require increasing use of ammonia or other non-carbonaceous ammonia derivatives.
- As ammonium nitrate (AN), which is very soluble in water. Because of the presence of the nitrate group, it is very fast acting and is immediately taken up by growing plants. To get the benefits of its immediate action, urea is often blended with ammonium nitrate. Ammonium nitrate is also widely used as a blasting explosive (as a slurry with diesel oil), because of this its production and distribution is highly regulated or banned in many jurisdictions.
- As complex fertilizers with other components providing the elements sulphur (ammonium sulphate, potassium sulphate), potassium (potassium sulphate) and phosphorus (diammonium phosphate).

Nitrogen fertilizers — ammonia, urea, AN — are particularly targeted to grasses and cereal crops — pasture, wheat, corn, rice and sugar. Complex fertilizers are generally used in horticulture producing vegetables and fruit crops. For explosives manufacture, ammonia is oxidised to produce nitric acid. This can then be used to produce ammonium nitrate or military explosives such as TNT or RDX.

Ammonia manufacture

Ammonia[12] is manufactured by the Bosch-Haber Process, which was first commercialised during World War I. The process has been much improved since then but generally remains the same. In this process, ammonia synthesis gas, a stoichiometric mixture of 3 parts hydrogen to 1 part nitrogen, is compressed and heated to about 20 MPa and 400°C and passed to a converter where the synthesis takes place using an iron catalyst, Figure 1.9:

$$3H_2 + N_2 = 2NH_3$$

After cooling the ammonia, it condenses and the liquid is separated and the unconverted synthesis gas is recycled. A purge from the separator vessel expels unwanted inert materials (such as methane and argon) which

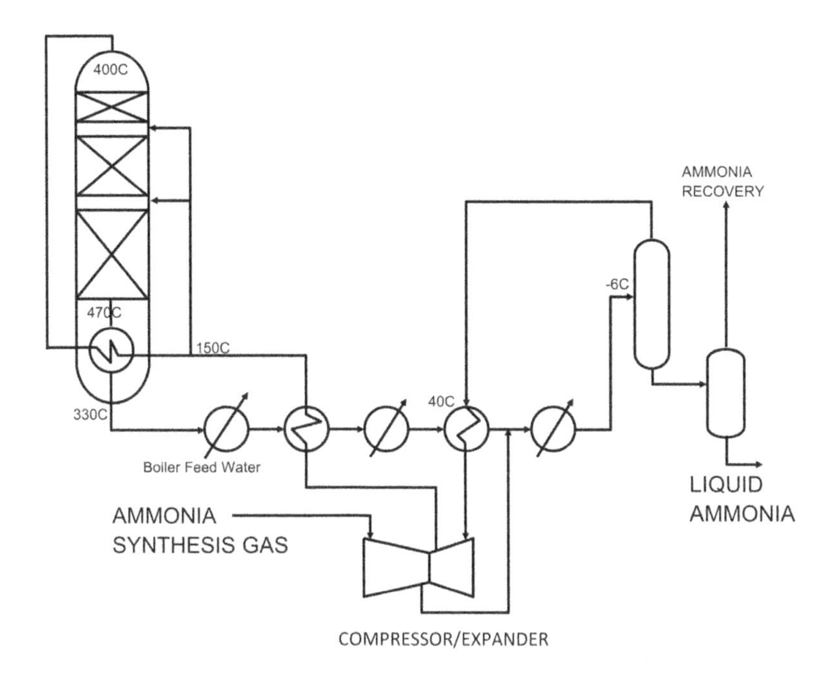

Fig. 1.9: Ammonia Synthesis.

[12] Jennings JR, Ward SA. (1989) Ammonia synthesis. In: Twigg MV (ed), *Catalyst Handbook,* 2nd ed. Wolfe Publishing Ltd, London.

build up in the recycle loop. The conversion is typically 12% so the recycle stream is large.

Importance of ammonia in the hydrogen economy

Ammonia contains 17.6% hydrogen by weight and since ammonia is a liquid that can be easily transported, the use of ammonia as a carrier for hydrogen is being widely promoted. This would avoid the high cost of other hydrogen transport systems such as liquefaction. The method has been researched for many years and is discussed more fully in later chapters.

For 150 Mt/y ammonia production, some 27 Mt/y hydrogens would be required from renewable sources if the industry is to move away from fossil fuel feedstocks.

The Use of Hydrogen in Reducing Metal Ores

Hydrogen, usually in the form of synthesis gas is used to reduce metal ores (oxides) to the metal. The most important is the reduction of iron ores (haematite (Fe_2O_3) or magnetite (Fe_3O_4)) for the manufacture of steel. The principal technology is the blast furnace which reduces the ore by means of coke made from coal. However, about 20% of the world steel is made by the so-called directly reduced iron (DRI) method using coal or increasingly natural gas which is converted into a hydrogen-rich synthesis gas. The basic process is very old (19th century) and was originally designed for coal reduction and was developed to incorporate natural gas as the reducing agent when natural gas became more generally available. It is practiced on a much smaller scale (typically 100–1000 kt/y) than that of a blast furnace, which is best for outputs over 2 Mt/y. The use of DRI is generally found in countries with modest steel demand. Over the years, there have been many variations on the process with many developed to accommodate the specific properties of a local iron ore.

When produced DRI can spontaneously ignite, to prevent this the product is often immediately passed to a briquetting plant. This eliminates the pyrophoric iron fines and facilitates the transport of DRI as a briquetted product, commonly called hot-briquetted iron (HBI), by road, rail or ship.

Most DRI operations require the iron ore as lump ore or as iron ore pellets produced in a pelletising operation. Because of the demand of DRI plants, iron ore lumps and pellets sell at a premium to iron ore fines. A few processes have been developed to use iron ore fines, which are cheaper than lump iron ore.

Direct Reduction of Iron Ore

Using haematite as the iron ore source, the main chemical processes are as follows:

$$Fe_2O_3 + 3H_2 = 2Fe + 3H_2O$$

and

$$Fe_2O_3 + 3CO = 2Fe + 3CO_2$$

Steel is an alloy of iron with up to 2% carbon. In the DRI process using natural gas, carbon is provided by the Boudouard reaction:

$$2CO = CO_2 + C$$

All of these reactions are reversible processes, so the extent of the reactions is dependent on the temperature, partial pressure (concentration) of reactants and products. The formation of DRI is favoured by high hydrogen content synthesis gas, commonly called reducing gas. Most processes remove produced water and some also remove carbon dioxide from the recycle gases.

Shaft reduction process

An example of a shaft reduction process is the Midrex™ Shaft Reduction Process[13] that has been developed specifically for natural gas reduction; The basic outline is illustrated in Figure 1.10.

[13] Midrex Corporation: www.midrex.com

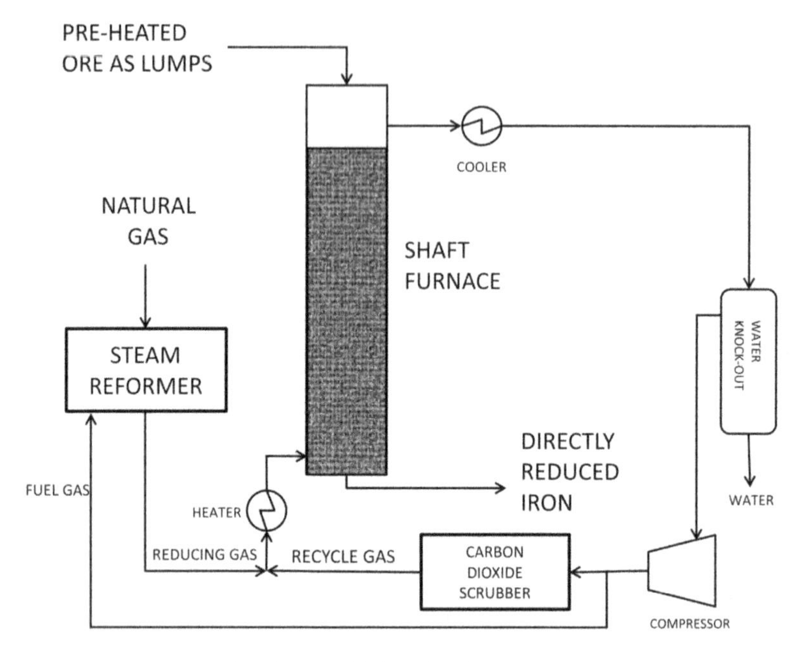

Fig. 1.10: Shaft reduction process.

In this process, a reducing gas is made from natural gas by a steam reformer.[14] The heated reducing gas is passed to the base of a shaft furnace and rises through the reactor reducing ore (as lump or pellets), which are preheated and added to the top section of the furnace. The exhaust gases leaving the top of the furnace are cooled and the produced water is condensed. The gases are recycled or purged to a fuel gas used to fire the reformer. The produced DRI is passed to a briquetting plant.

Iron ore fines reduction process

An example of a process designed for the reduction of iron ore fines is the FINMET™ process.[15] This process is a development of the earlier FIOR process. This process was developed by Exxon in the 1970s and was first commercialised in Venezuela. In the 1990s, BHP built a large version of

[14]The operation of a steam reformer is discussed in more detail in the next chapter.

[15]Hasssan A. Whipp R. The Finmet process — an operational and technical update, *Stahl und Eisen*, **124**(4): 53 (2004).

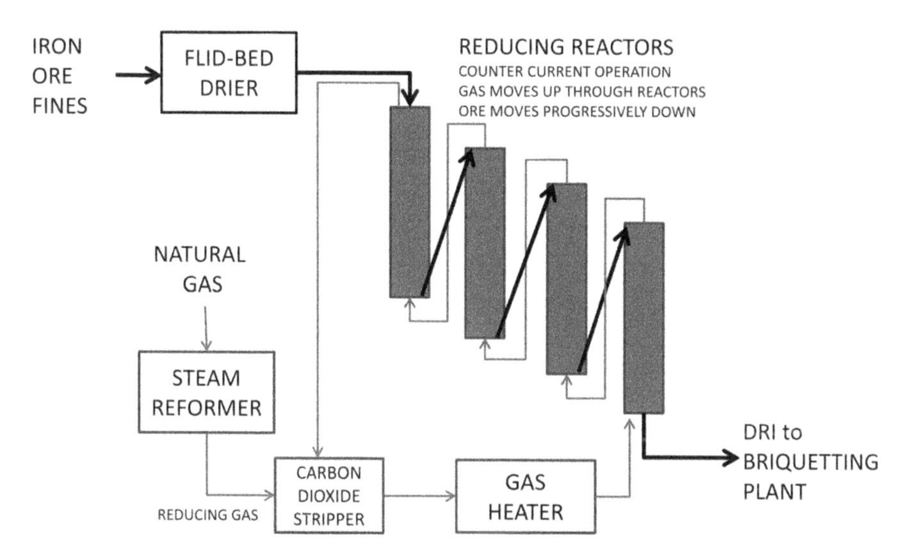

Fig. 1.11: DRI by the FINMET™ process.

the plant processing 2 Mt/y of iron ore fines into HBI at Port Hedland in North West Australia. It had several operational problems and was closed in 2007.

In this process, the iron ore fines are dried in a fluid-bed drier and passed to the first of a bank of fluidised-bed reactors: Figure 1.11. The ore is progressively passed down the line of reactors where the ore is progressively reduced. Natural gas is passed to a steam reformer to produce the reducing gas which is then passed to a carbon dioxide removal unit, this unit also removes excess water. The gas is heated and passed to the base of the last reactor where the gas ascends against the descending iron fines. The product is passed to a briquetting plant.

Role of DRI in the hydrogen economy

There is increasing interest in using hydrogen as the only reducing agent in the manufacture of steel. Clearly, DRI processes based on natural gas would be candidates for conversion from a fossil fuel sourced reducing agent to a renewable hydrogen or zero carbon emission source. Should carbon capture and storage be a viable option, then the present technology for DRI could be used if it is equipped with a carbon dioxide removal system. Fortunately, unlike air combustion of fossil fuels, carbon dioxide emissions from DRI facilities should be easy to remove producing a relatively pure carbon dioxide stream that could be compressed and geosequestrated. Another approach being pursued by some companies is to substitute some of the natural gas with renewable hydrogen with the hope of totally replacing the natural gas in the future.[16]

In 2017, world steel production was over 1700 Mt. If all this were to be produced by renewable hydrogen, the hydrogen requirement would be over 180 Mt/y.

Observations

Large quantities of hydrogen based on fossil fuel feedstock are used in the process industries. Many countries and jurisdictions have targeted net-zero carbon emission by 2050. The current demand for hydrogen in the petroleum and chemical industries is well over 50 Mt/y. Replacing present technology for the production of steel would require over 180 Mt/y hydrogen. If the whole of the crude oil stream is to be replaced (90 million barrels per day, equal to about 186 EJ/y), this would require an annual hydrogen production from zero carbon or renewable sources of over 1500 Mt/y.

This is a lot of hydrogen.

[16] See "MidrexH2" at www.midrex.com

CHAPTER 2

METHODS OF HYDROGEN PRODUCTION

In Chapter 1, we described the current large demand for hydrogen and this demand will only grow as a 'Hydrogen Economy' develops. Terrestrial hydrogen is sourced from either carbonaceous materials or water. Current manufacture is dominated by production from fossil fuels — coal, oil or gas. For this, to continue in a carbon-constrained world then carbon capture and storage is a necessity. If the use of fossil fuels is to be eliminated, then the major source will be water alone. This chapter discusses the production of hydrogen from carbonaceous sources and water.

The Production of Hydrogen from Coal

When measured on a dry and ash-free basis, carbon is the dominant element in coal ranging from about 60% to over 80% of the mass. The hydrogen content of coal is very low — typically about 5% by mass. Gaseous hydrogen is made from coal by a combination of two processes — gasification and water-gas shift. The gasification reaction is a partial oxidation process that is applicable to the combustion of carbonaceous solids. It was originally developed for coal but with adaption is applicable to any carbonaceous solid including biomass and municipal waste streams (MWS).

Gasification

In this process, the solid feed is burned in a stream of oxygen (or in some cases air) and steam. The process is conducted in a gasifier of which there are many types. The chemistry of combustion is complex and the main processes involved within the gasifier are:

Solid gas reactions

Combustion: This reaction is highly exothermic and provides the heat for the process.

$$\mathbf{C + O_2 = CO_2}$$

Steam — Carbon Reaction: This reaction converts water (steam) into hydrogen.

$$\mathbf{C + H_2O = CO + H_2}$$

Hydro-gasification: This reaction consumes hydrogen and forms methane.

$$\mathbf{C + 2H_2 = CH_4}$$

The Boudouard reaction: This converts carbon dioxide into carbon monoxide. This reaction is reversible and can cause the deposition of carbon in the process plant remote from the gasifier.

$$\mathbf{C + CO_2 = 2CO}$$

Gas phase reactions

The Water-Gas-Shift (WGS) Reaction: A key reaction for making hydrogen from solids. It converts carbon monoxide and steam into carbon dioxide and hydrogen. This reaction is reversible.

$$\mathbf{CO + H_2O = CO_2 + H_2}$$

Methanation: This process consumes hydrogen to produce methane. This reaction is reversible, with the reverse process being steam methane reforming, which is discussed later.

$$\mathbf{CO + 3H_2 = CH_4 + H_2O}$$

Process optimisation

The extent of all of these processes is determined by the thermodynamics pertaining at the operational temperature and pressure of the gasifier for any given feedstock, oxidant (oxygen or air) and steam injection. The composition of the resulting synthesis gas can be estimated, that is the relative amounts of hydrogen, carbon monoxide, carbon dioxide, water and methane determined by solving the pertinent thermodynamic equations.[1] For maximum process efficiency, the quantum of carbon monoxide plus hydrogen in the synthesis gas should be maximised. The process is optimised by changing the composition of the input stream and the operating temperature and pressure of the gasifier. For example, increasing the oxygen input will increase the gasifier operating temperature and this will change the relative amounts of the products in the synthesis gas. Different types of gasifiers have differing optimum operating conditions.

Pyrolysis

In pyrolysis processes, large molecules are cracked (pyrolysed) or hydropyrolysed into smaller molecules. The resultant products can undergo gasification to the desired products, but if this does not occur, because, for instance the temperature is too low pyrolysis tars and liquids are produced. These foul downstream plants, if not removed from the product, synthesis gas stream.

The production of tars and liquids should be avoided if possible. The process is associated with gasification at relatively low temperatures; 600°C or so as opposed to temperatures in excess of 900°C or higher. The problem is a particular issue with feed containing high levels of water and/or oxygen such as lignite, biomass and MWS.

Increasing the gasification temperature by adding more oxygen can minimise pyrolysis products but excess oxygen results in the preferential burning of hydrogen rather than carbon monoxide:

$$\mathbf{2H_2 + O_2 = 2H_2O}$$

[1] Montgomery CW, Weinberger EB, Hoffman DS. (1948) Thermodynamics and stoichiometry of synthesis gas production. *Ind Eng Chem* **40**(4): 601–607.

Gasifier Types

Individual gasifiers types have been developed for specific feeds, some have been adapted so that other feeds can be accommodated. Four main types of gasifiers are discussed:

Moving bed gasifier

This gasifier type is often referred to as the Lurgi gasifier. It was developed for black coal initially using air as the oxidant. The main aim of this was to produce an ammonia synthesis gas. There are many of these gasifiers in operation. An advanced gasifier, which uses oxygen and steam as the oxidant, was developed by British Gas (BG/Lurgi gasifier). This increased the gasifier operating temperature and pressure and made it suitable for poorer quality coals such as lignite. The gasifier waste was produced as a molten slag rather than ash. Figure 2.1 shows the general layout.

The gasifier is well proven commercially with many hundreds in use. It comprises a pressure vessel and can operate up to about 3 MPa. These gasifiers are relatively small and large facilities generally employ banks of 50 or more. The basic design is easily adapted for different feedstocks including biomass.

Coal (as lumps — typically 1 inch) enters the top of the gasifier through a series of hoppers. The coal falls onto an incandescent bed of burning coal and progressively moves down the bed (hence moving-bed gasifier). Steam and oxygen enter the bottom of the gasifier and pass upwards through the bed. At the base of the bed is a system of rotating grates that holds up the bed and allows ash or slag to fall to the bottom of the gasifier for removal and disposal.

The main issues with this gasifier type are the small size for large facilities and relatively low operating pressure. The main operational issue is that fresh feed coal (or other feed) falling on onto the hot coal bed undergoes pyrolysis and the resulting tar and liquid products contaminate the synthesis gas stream. The synthesis gas is typically passed to a cold methanol wash,[2] which efficiently removes the tar and liquid products.

---

[2] Linde Engineering: Rectisol Wash™.

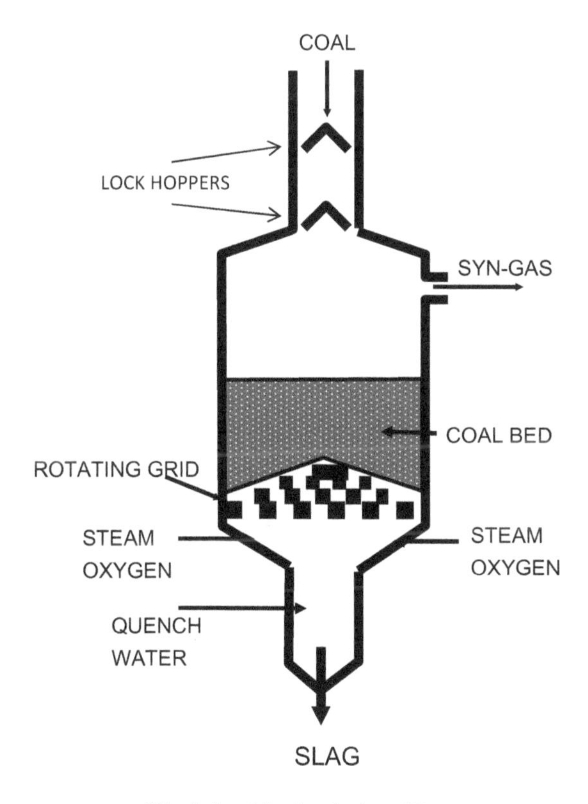

Fig. 2.1: Moving bed gasifier.

It can also be used to remove carbon dioxide and hydrogen sulphide from the gas.[3]

Fluid-bed gasifier

This was originally developed as the Winkler gasifier over 100 years ago and has been developed further by KBR[4] (KRW gasifier) to operate at high temperature with a wide range of feedstock; Figure 2.2. The gasifier

---

[3] Ranke G, Mohr V. (1985) Rectisol wash — new developments for acid gas removal from synthesis gas. In: SA Newman (ed), *Acid and Sour Gas Treating Processes*. Gulf Publishing Company.

[4] Kellogg, Brown and Root.

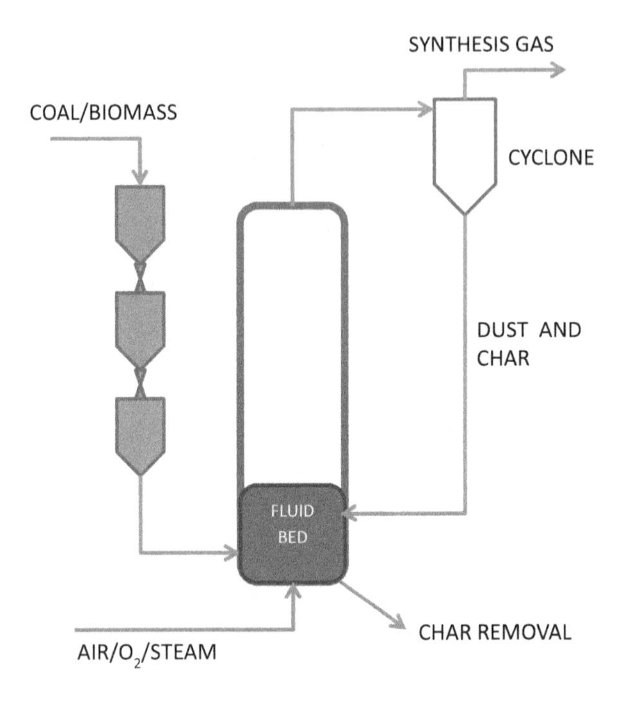

Fig. 2.2: Fluid-bed gasifier.

comprises a large pressure vessel, often with a double skin containing water that cools the gasifier and generates steam. The feed, coal or biomass is pre-dried and reduced to fines and charged into a series of lock hoppers before entering the gasifier at the base of the reactor. The feed is fluidised by a stream of oxidant (air or oxygen) and steam. The fluidisation ensures a homogeneous temperature profile that reduces the amount of pyrolysis products. The fluid bed lies at the bottom of the reactor and a large free space above the bed allows solids to disengage from the gas stream and allowing the gas phase reactions to go to completion. The ash, as an agglomerate, is removed from the bottom of the reactor through a hopper system (not shown).

The hot gas products are passed to a cyclone that removes dust and char recycling these to the fluid-bed from the product synthesis gas.

There are several of these type of fluid-bed gasifiers in operation (e.g. U-Gas™ type[5]). They have lower capital cost than other types of gasifiers

[5]The U-Gas Gasifier was developed by the Gas Technology Institute: www.gti.energy.

such as the entrained-bed type. The main problem is that, as a consequence of fluidisation, there is a finite chance that some entering feed exits in the char (ash) removal system. This contaminates the char with typically 1% carbon. This represents lost product. The contamination can degrade the use of the ash agglomerate for some uses.

Entrained-bed gasifiers

This is the most efficient gasifier type. The origin is in the development of the Koppers-Totzek gasifier in the 1950s. This technology is the basis of the Shell Gasifier and the Uhde Prenflo Gasifier; Figure 2.3.

The gasifier comprises a large double-walled pressure vessel operating up to 5 MPa. The coal is burned as fine power that is injected into the gasifier by means of a nitrogen stream and mixed with oxygen in burners

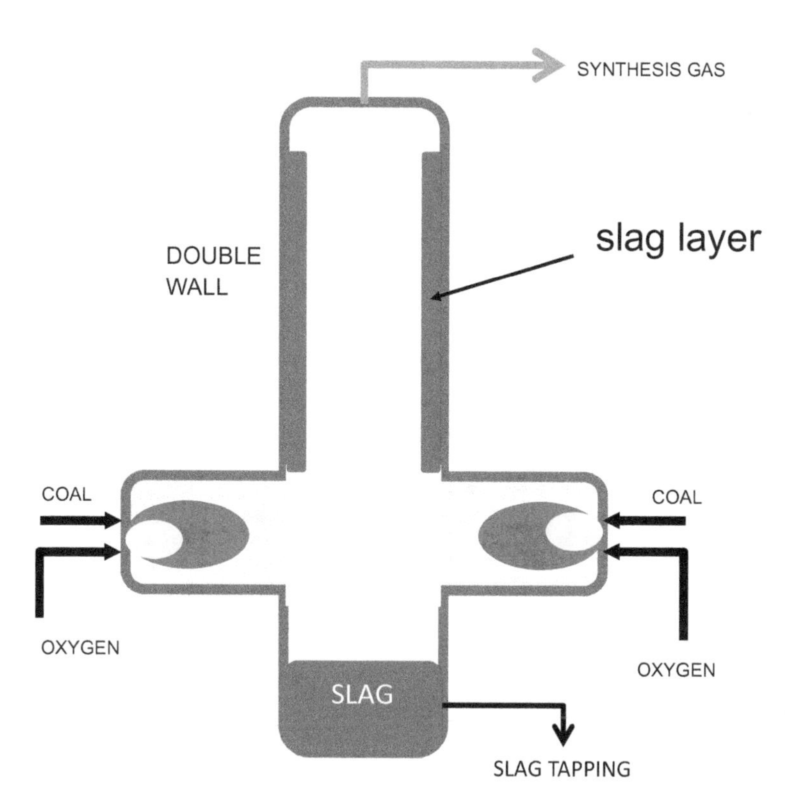

Fig. 2.3: Entrained-bed gasifier.

near the bottom of the gasifier. Complete gasification of the coal occurs at a high operating temperature, higher than the ash fusion point (typically about 1400°C). If the ash fusion temperature is higher than the gasifier operation, fluxes such as limestone are added to the coal mix. The molten slag is carried upwards within the gasifier with the hot gases and deposits on the walls of the gasifier from where it falls to the floor of the unit and is tapped out of the gasifier for disposal. Synthesis gas passes out of the top of the gasifier to a filter and waste heat boiler for energy recovery as high-pressure steam. Because of the high efficiency of the process, the main contaminant of the synthesis gas is ash dust and this can be removed by means of candle filters as opposed to the higher cost liquid scrubbing operations. Because the raw synthesis gas contains no pyrolysis products, carbon dioxide and hydrogen sulphide can be removed with lower cost amine scrubber units.

Texaco slurry entrained-bed gasifier (GE gasifier)

This type of gasifier introduces the feed into the gasifier as a water, or sometimes hydrocarbon, slurry. The gasifier is designed to operate on low moisture content black coal. There are many variations with some designed for the gasification of heavy oils, naphtha or natural gas. In essence, the coal is dried and finely ground and is fed to a burner system as a water slurry to the top of the gasifier. The burner results in a downward flame; Figure 2.4.

The gasifier is a large vessel that is lined with firebrick. The burner temperature is very high and is higher than the ash fusion point of the coal so forms a slag that is collected at the bottom of the vessel from where it is removed. The hot synthesis gas is passed to a waste heat boiler and then a water scrubber that removes soot from the product gas. This soot can be recycled to the gasifier.

There are several variants on this type. Some employ a water-cooled jacket around the main reactor and some have heat exchange equipment within the gasifier. The capital cost of the gasifier can be reduced by replacing the downstream heat exchange and water scrubber with a water quench.

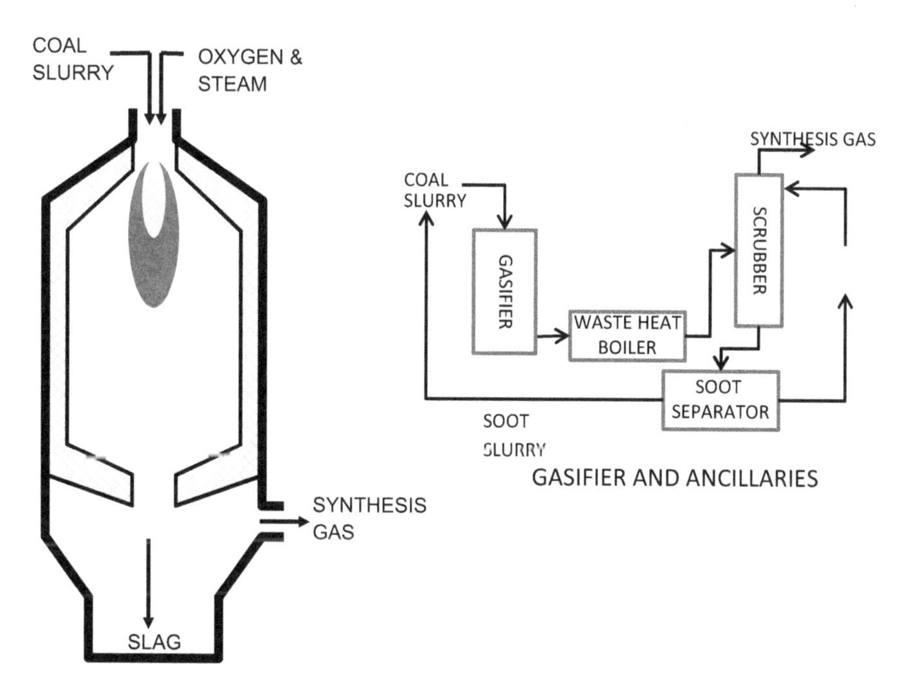

Fig. 2.4: Texaco (GE energy) gasifier.

There are many of these types of gasifiers in operation (particularly using petroleum coke or liquid feedstocks), with operating pressures up to 5 MPa. Because of the water slurry injection system, this type of gasifier is generally not suitable for high moisture content feedstocks such as lignite, biomass or MWS.

Comparative performance of gasifiers

Table 2.1 illustrates the relative performance of a range of gasifiers on the same coal (Illinois No. 6). A proximate and ultimate analysis of this coal is given in Table 2.2. As will be discussed later, the production of hydrogen is completed by the WGS process that converts every mole of carbon monoxide present in the synthesis gas into a mole of hydrogen, so that the critical factor is the total CO + H_2 produced; see bottom line of Table 2.1. This indicates the superiority of the entrained-bed gasifier in maximising

Table 2.1: Relative performance of various gasifier types using Illinois No. 6 coal.[6]

		Lurgi	BG/Lurgi	KRW	Texaco	Shell
Type of bed		Moving	Moving	Fluid	Entrained	Entrained
Pressure	MPa	0.1	2.82	2.82	4.22	2.46
Ash type		ash	slag	agglom.	slag	slag
H_2	Vol%	52.2	26.4	27.7	30.3	26.7
CO	Vol%	29.5	45.8	54.6	39.4	63.1
CO_2	Vol%	5.6	2.9	4.7	10.8	1.5
CH_4	Vol%	4.4	3.8	5.8	0.1	0.03
Other hydrocarbons	Vol%	0.3	0.2	0.01	0	0
H_2S	Vol%	0.9	1.0	1.3	1	1.3
Ratio H_2S/COS		20/1	11/1	9/1	42/1	9/1
N_2 + A	Vol%	1.5	3.3	1.7	1.6	5.2
H_2O	Vol%	5.1	16.3	4.4	16.5	2.0
NH_3 + HCN	Vol%	0.5	0.2	0.8	0.1	0.02
Gas totals	Vol%	100	99.9	101.01	99.8	99.85
CO + H_2	Vol%	81.7	72.2	82.3	69.7	89.8
Dry basis total	Vol%	94.9	83.6	96.61	83.3	97.85
CO + H_2 dry gas	Vol%	86.09	86.36	85.19	83.67	91.77

this parameter and minimising unwanted by-products — carbon dioxide, and higher hydrocarbons. The atmosphere in the gasifier is reducing in nature and as a consequence, the sulphur in the coal is primarily reduced to hydrogen sulphide; there is some carbonyl sulphide formed.

Although the most efficient gasifiers have different outcomes in terms of the stoichiometric ratio of the synthesis gas (stoichiometric ratio (SR) < 1), they broadly reflect the molar composition of C/H in the feed. The exception is the relatively low efficiency, low-pressure Lurgi gasifier that sees an SR of >1.

For maximum hydrogen production, every mole of carbon monoxide is shifted to a mole of hydrogen by the WGS process discussed below.

[6] Bhandarkar PG. Gasification Overview Focus — India, Hydrocarbon Asia, pp. 46–51, November/December 2001 (www.hcasia.safan.com/mag/dec01/t46.pdf).

Table 2.2: Ultimate and proximate analysis of Illinois No 6.

Ultimate Analysis (Dry Ash Free, DAF)			**Mol%**
Carbon	wt%	78.10%	64.16
Hydrogen	wt%	5.50%	27.11
Oxygen	wt%	10.90%	6.72
Nitrogen	wt%	1.2%	0.70
Sulphur	wt%	4.30%	1.32
		100.00%	100.01
Ash (as received)	wt%	12.0%	
Moisture (as received)	wt%	6.5%	
As received Basis			
Carbon	wt%	65.91%	
Hydrogen	wt%	4.64%	
Oxygen	wt%	9.20%	
Nitrogen	wt%	1.01%	
Sulphur	wt%	3.63%	
Ash (as received)	wt%	10.13%	
Moisture (as received)	wt%	5.49%	
		100.01%	
LHV (as received)	GJ/t	25.80	
HHV (as received)	GJ/t	25.90	
LHV (Dry Ash Free [DAF])	GJ/t	30.57	
HHV (DAF)	GJ/t	30.69	

Before discussing WGS, the production of hydrogen from natural gas will be discussed.

The Production of Hydrogen from Natural Gas

Natural gas is substantially methane. Unlike coal which contains little hydrogen, methane contains 25% by weight hydrogen (molar H/C is four). When natural gas became more widely available in developed economies, particularly during the 1960s, the production of hydrogen-rich streams from natural gas displaced older technology based on coal. This route is now the dominant technology in The Organisation for Economic Co-operation and Development (OECD) nations.[7]

Methane is a relatively inert compound with four strong hydrogen to carbon bonds. It is difficult to convert into other materials. However, in the presence of oxygen (as gas) or water at elevated temperatures, it is thermodynamically unstable relative to synthesis gas. There are two main approaches to synthesis gas production — steam reforming and partial oxidation and several hybrid approaches such as auto-thermal reforming.

Steam Reforming

In steam reforming, methane (or higher hydrocarbons) is mixed with excess steam and passed over a nickel catalyst at about 850°C. At this temperature, in a strongly endothermic reaction, the methane is reformed into synthesis gas:

$$\mathbf{(1+x)CH_4 + H_2O = CO + 3H_2 + xCH_4} \quad \Delta H^\circ_{298}\ +206\text{kJ/mol}$$

The molar ratio of hydrogen to carbon monoxide (the stoichiometric ratio) is quite high at 3/1. In practice, it is somewhat higher because at the reforming temperature, the WGS reaction converts some of the carbon monoxide to carbon dioxide and water:

$$\mathbf{CO + H_2O = CO_2 + H_2} \quad \Delta H^\circ_{298}\ -41\text{kJ/mol}$$

The resulting SR ratio of the reformed gas is about 4. This can be altered somewhat for different uses, for instance, reducing gas for iron ore reduction is best with a high hydrogen content in the gas. The final composition of the synthesis gas is a function of the reforming temperature, pressure (up to 3 MPa) and amount of excess water present. This is varied according to the final use.

[7] The conversion of natural gas and the downstream processes water-gas-shift and methanation are extensively discussed in *Catalyst Handbook* (2nd Edition, M.V. Twigg ed.), Wolfe Publishing 1989. Some of the many variants of synthesis gas (syngas) manufacture are discussed by Rostrup-Nielsen JR, Christiansen LJ. (2011) *Concepts in Syngas Manufacture, Catalytic Science Series — Vol. 10.* ICP Press.

The term **x** in the reforming reaction is known as the methane slip. This is quite high for steam reforming (several percent of the synthesis gas). This is acceptable for some duties such as the production of ammonia synthesis gas when a secondary reformer burns this methane in air and thereby introduces nitrogen in into the synthesis gas stream.

Steam reforming is commercially mature and well developed. There are many variations of the basic process and many thousands in operation. Figure 2.5 illustrates the operation of a multi-tubular reformer that is a common approach for large-scale methane reforming.

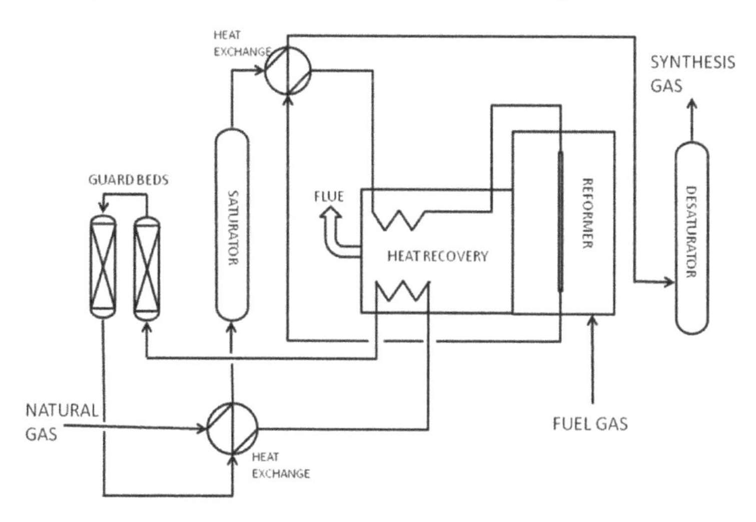

Fig. 2.5: Multi-tubular steam reformer.

The reformer resembles a large box like structure with the reformer tubes (each about 10 m long and 10 cm in diameter packed with catalyst) hanging from the roof area; a large reformer can have more than 3000 tubes. These are heated by gas burned from the top, bottom or sides (this varies according to the specific technology). Hot gases from the reformer section are passed to large heat recovery box containing piping for pre-heating the reformer gases and raising process steam prior to passing to a flue.

Natural gas feed enters the system, and is passed to guard beds to remove all of the sulphur in the feed stream that would otherwise poison the catalyst. The gas is then passed to a saturator where the warm gas evaporates water for the process; many processes inject steam rather than using a saturator. Typically, between 2–6 moles of water are used per mole of methane fed.

The gases are heated further and passed to the reformer tubes. The reformed gases are then cooled and excess water removed to produce the raw synthesis gas. The synthesis gas comprises carbon monoxide, carbon dioxide, hydrogen, methane and saturated water. There are no contaminants such as dust, ash or carbon (soot).

Despite the size of the units, steam reformers are relatively easy to construct.

Maximising Hydrogen Yield

Steam reformers can give a high hydrogen yield from natural gas. This is illustrated by the manufacture of reducing gas for the production of DRI (see above). For this duty, a steam reformer is operated at relatively high temperatures (800–1000°C), low pressure (0.3–1 MPa) and with a low steam/carbon ration of about 2.2 (Table 2.3).

Table 2.3: Reducing gas — maximising hydrogen in a steam reformer.[8]

	Inlet	**Exit**
Pressure (MPa)	1.07	0.7
Temperature(°C)	510	830
Methane (vol%)	85.5	4.0
Ethane (vol%)	9.1	
Propane (vol%)	0.6	
Carbon dioxide (vol%)	3.7	7.0
Nitrogen (vol%)	1.1	1.0
Carbon monoxide (vol%)		16.0
Hydrogen (vol%)		72.0

[8] Ridler DE, Twigg MV. *Steam Reforming* In: MV Twigg (ed), Catalyst Handbook, 2nd ed. Wolfe Publishing, 1989.

Partial Oxidation

Partial oxidation is similar to gasification and is sometimes referred to as such. In this process, natural gas is burned in a restricted quantity of oxygen with steam added as a flame moderator. The equation for the exothermic process is

$$\mathbf{(1 + x)CH_4 + 0.5O_2 = CO + 2H_2 + xCH_4} \quad \Delta H^\circ_{298}{-36kJ/mol}$$

In contrast to steam reforming, the reaction is strongly exothermic and no additional heat input is required. The amount of methane slip (**x**) is very low, a few hundred parts per million (ppm), and measurement of this factor is used to ensure the system does not form an explosive mixture within the unit. The operating temperature is very high, typically 1500°C or more. At this temperature, the reaction thermodynamics do not favour the formation of carbon dioxide and this is limited to about 2% of the synthesis gas. There is some carbon (soot) produced in the process (mainly from the presence of higher hydrocarbons) that has to be scrubbed out of the product synthesis gas.

In practice, the SR of the synthesis gas is <2 due to some combustion of the produced hydrogen in preference to carbon monoxide.[9]

The partial oxidation gasifier comprises a pressure vessel lined with firebrick with a combustion lance in the top of the unit to which the natural gas, oxygen and steam are fed; Figure 2.6. Unlike the steam reformer, sulphur removal is not required, the sulphur compounds are reduced to hydrogen sulphide. Partial oxidation units can operate to 10 MPa, which is an advantage for some downstream uses. The hot gases are passed to a waste heat boiler that generates high pressure stream. The gases are further cooled and passed to a water scrubber (some systems have a naphtha scrubber) to produce the raw synthesis gas. Soot from the scrubber can be recycled to the gasifier.

Partial oxidation gasifiers are much smaller than steam reformers and are often used when space is a premium. They are apparently of lower cost than steam reformers but the capital cost often ignores the cost of the air

[9] Note this is higher than coal gasification that is generally gives an SR below unity.

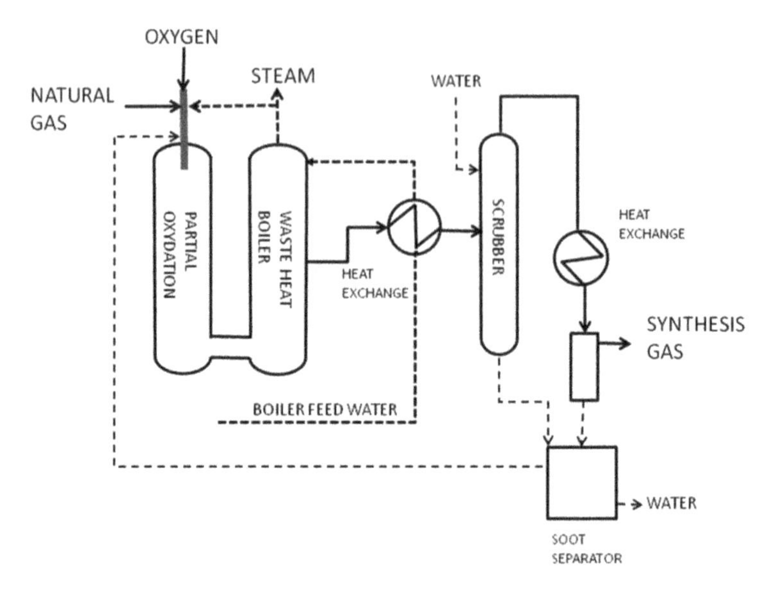

Fig. 2.6: Natural gas partial oxidation.

separation units (ASU), which provides the oxygen. Including the ASU into the capital makes the partial oxidation unit a similar cost to that of the steam reformer.

Auto-Thermal Reforming

The auto-thermal reformer marries together a steam reformer and a partial oxidation unit. The process aims to produce synthesis gas with an SR of 2–2.5, which suits many downstream process operations (methanol, Fischer-Tropsch Process) and seeks to avoid the WGS reactor to adjust the final SR thus lowering the overall capital; Figure 2.7 illustrates the layout.

The auto-thermal reactor is a very large unit separated into two sections,[10] the top being a partial oxidation unit and the lower section a packed-bed steam reformer. Because of the use of steam reformer

[10]Again there are several variants on its technology, some offering a compact reactor design.

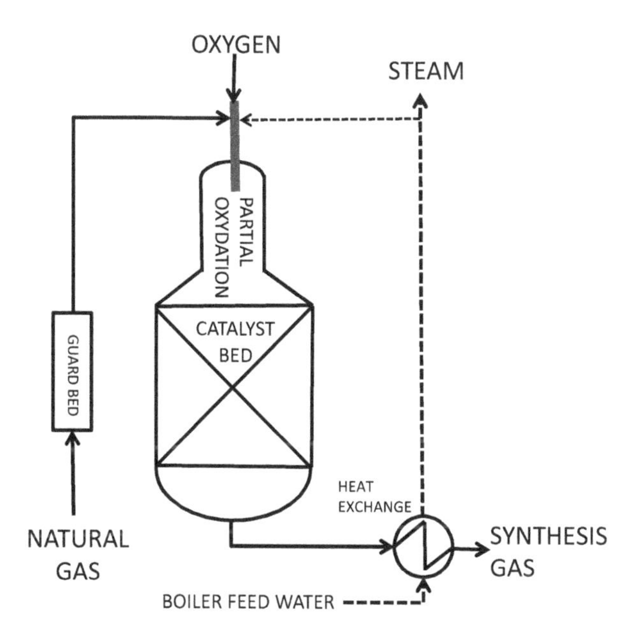

Fig. 2.7: Auto-thermal reforming.

catalysts, the natural gas is pre-treated to remove sulphur prior to passing to the combustion lance at the top of the unit where it is mixed with oxygen from the ASU and steam. The unit is designed so that the heat of the partial oxidation section is sufficient for the heat input into reforming section. Less oxygen is used so there is a significant amount of methane passing to the reformer section that converts this into synthesis gas. The hot gases are cooled and steam generated and the synthesis gas is passed to downstream process units.

The Production of Pure Hydrogen

The above sections concern the production of synthesis gas, a mixture of carbon oxides, mainly carbon monoxide, and hydrogen. To produce pure hydrogen, it is necessary to cause more hydrogen to be formed out of the carbon monoxide by the WGS process and then to purify the resultant crude hydrogen stream.

The WGS Process

Water-gas

The WGS was an early process for the production of hydrogen-rich streams from coal for the production of ammonia for explosives and fertiliser. The process dates from the coal industry prior to the introduction of gasifiers. In these early days, synthesis gas was produced by a batch process in two vessels charged with coke; Figure 2.8.

In the first vessel, air was used to burn the coke to an incandescent temperature. The main reactions of combustion are:
Oxidation

$$\mathbf{C + O_2 = CO_2} \quad \Delta H^{\circ}_{298} -394 \text{kJ/mol}$$

and the reverse Boudouard reaction to produce carbon monoxide.

$$\mathbf{C + CO_2 = 2CO} \quad \Delta H^{\circ}_{298} +172 \text{kJ/mol}$$

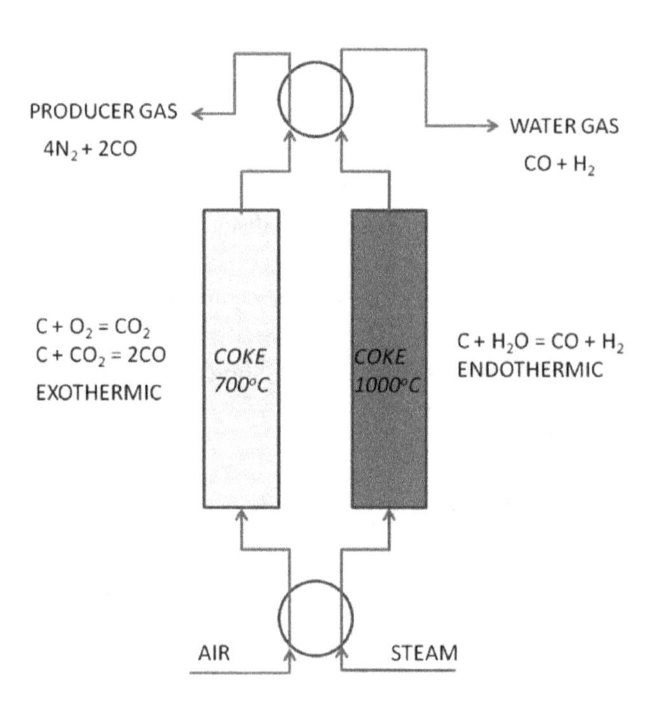

Fig. 2.8: The production of water-gas.

With the nitrogen from the air, this produces 'Producer Gas,' which although of low calorific value can be used as a fuel to help power the system.

$$2C + O_2 + 4N_2 = 2CO + 4N_2$$

When the reactor has reached a suitable temperature, a series of switch valves (not shown) pass steam through the vessel where the coke is converted into synthesis gas, known as water-gas.

$$C + H_2O = CO + H_2 \qquad \Delta H^{\circ}_{298} + 131 kJ/mol$$

This reaction is endothermic and cools the coke bed. The steam is then replaced by air that reheats the bed to the required temperature. The two beds worked in tandem to produce a continuous stream of producer-gas and water-gas. This process has been replaced by the more efficient gasifier processes discussed above.

A gas suitable for the production of ammonia was obtained by judicious mixing of producer gas and water gas.

Water-gas-shift

As seen above, the gasification of coal produces a synthesis gas with low hydrogen content. This is corrected by the WGS process, which converts carbon monoxide and water (steam) into carbon dioxide and hydrogen:

$$CO + H_2O = CO_2 + H_2 \qquad \Delta H^{\circ}_{298} - 41.1 kJ/mol$$

The reaction is reversible, with low temperatures favouring the forward reaction as written above. In the process, excess steam is added to the synthesis gas and the gas passed over an iron–chromium catalyst at temperatures in excess of 400°C. This is known as the high temperature WGS (HTWGS) and it converts synthesis gas with a carbon monoxide content over 10% to a synthesis gas with a carbon monoxide content of below 3%. This limit is due to the thermodynamic constraints set by the HTWGS temperature. The HTWGS catalyst system is very robust to impurities such as sulphur compounds and these do not have to be removed for the raw synthesis gas prior to HTWGS.

If the carbon monoxide of the synthesis gas is to be reduced further to obtain a purer hydrogen stream, then a low temperature WGS (LTWGS) process is required. This takes place over a copper catalyst at 200°C and

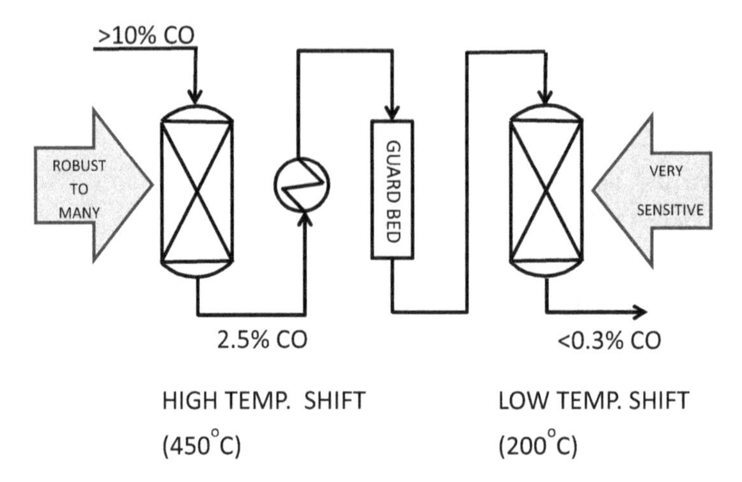

Fig. 2.9: The water-gas-shift process.

this reduces the carbon monoxide in the gas stream to <0.3%, i.e. generating a hydrogen stream of >99% purity relative to carbon monoxide. The LTWGS catalyst is very sensitive to impurities in the stream including sulphur and this has to be removed prior to the LTWGS reactor. This duty is typically provided by a zinc oxide guard bed; Figure 2.9.

To obtain a product hydrogen stream, the carbon dioxide resulting from the WGS is stripped out of the system. There are many commercial processes for this. They typically involve contacting a carbon dioxide-rich gas with a solvent that extracts the carbon dioxide, the carbon dioxide loaded solvent is passed to another column where the fresh solvent is regenerated by boiling-off the carbon dioxide; a typical absorber-stripper system is illustrated in Figure 2.10.

The carbon dioxide stripping system[11] can also be used to remove sulphur as hydrogen sulphide if this is present. One important point to note is that the carbon dioxide (possibly along with hydrogen sulphide) is obtained as a pure stream from the regenerator. This is generally discharged to atmosphere. For carbon capture and storage, this stream requires cooling and compression to produce liquid (or dense phase) carbon dioxide.

[11]The many approaches to carbon dioxide and hydrogen sulphide removal from gas streams is discussed by Newman SA. ed. (1985) *Acid and Sour Gas Treating Processes*. Gulf Publishing.

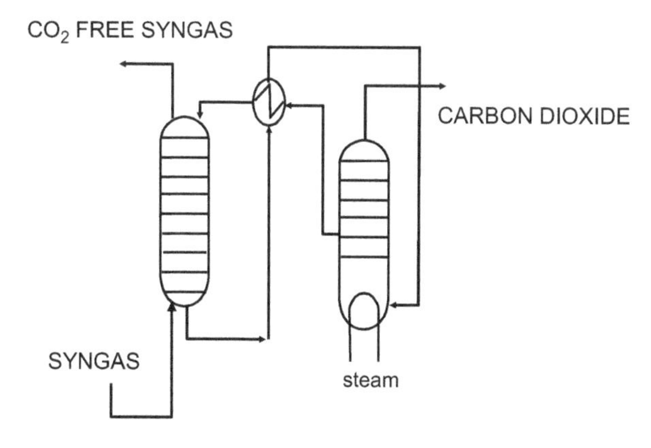

Fig. 2.10: Carbon dioxide stripping.

The result of WGS and carbon dioxide stripping is a hydrogen stream of about 95% purity. Also present is methane from the gasifier operation and traces of carbon oxides. Methane is relatively inert and generally does not present a problem to downstream hydrogen processes but the carbon oxides cause poisoning or fouling of many downstream catalysts.

Hydrogen Purification

Some uses of hydrogen may be tolerant to the remaining impurities; e.g. electricity production from high temperature solid oxide fuel cells. However, if these remaining impurities are an issue for downstream operations, then the hydrogen needs to be purified further.

Methanation

The first approach that is well known for the production of ammonia is to convert the remaining carbon oxides to methane which is inert in the ammonia synthesis process. The process outline is shown in Figure 2.11.

Hydrogen that after passing through HTWGS, LTWGS and carbon dioxide stripping and with a carbon oxide content of less than 1% is heated and passed to a methanation reactor at about 350°C. This removes the carbon oxides by sacrificing some of the hydrogen to produce methane.

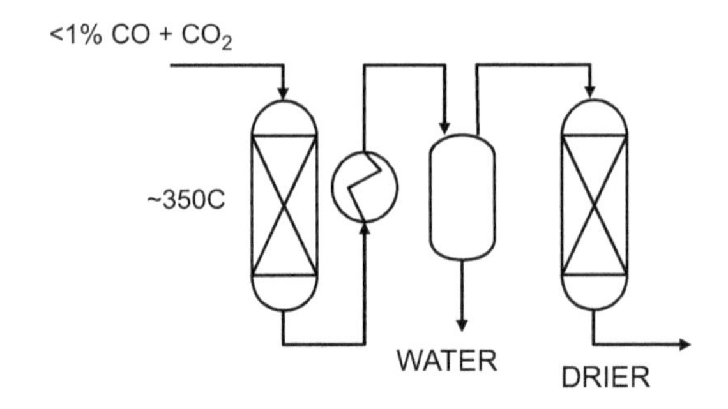

Fig. 2.11: Methanation.

$$\mathbf{CO + 3H_2 = CH_4 + H_2O}$$

$$\mathbf{CO_2 + 4H_2 = CH_4 + 2H_2O}$$

Cooling the effluent removes most of the produced water, the hydrogen product is finally produced after passing through a drier. The resulting hydrogen stream containing some methane is suitable for most downstream uses.

Separation

If methane in the stream is a problem, then this can be removed by one of three processes:

Pressure-swing adsorption (PSA)

Pressure-swing adsorption (PSA) is widely used to separate hydrogen from streams contaminated with one or several impurities. In essence, the gas to be treated at pressure (usually 1 MPa–4 MPa) is passed to an adsorber column containing a molecular sieve adsorbent material (e.g. ion exchange zeolite A or X); Figure 2.12.

The hydrogen passes through the column and the impurities with boiling points higher than hydrogen (all except helium) are adsorbed. When the sieve becomes saturated, the feed stream is switched to another column and the saturated column is regenerated. In PSA, this is accomplished by reducing the pressure or applying a vacuum and the contaminants are

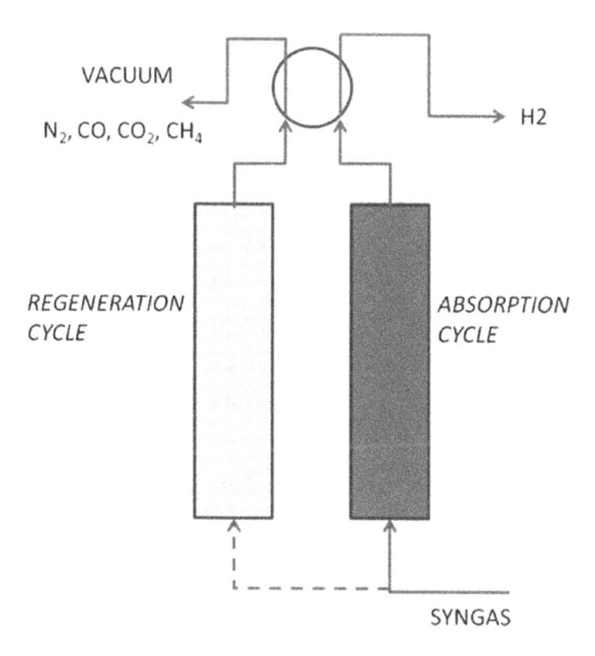

Fig. 2.12: Pressure-swing absorption.

released. In some systems, regeneration is assisted by heating the beds or using steam to drive off the adsorbed gases; these systems are generally referred to temperature swing adsorption (TSA) processes. In practice, a battery of adsorbent beds is used with a complex of valve switching between adsorbing and regenerating beds.

In recent years, the PSA system has been developed to reduce the overall capital cost of hydrogen production by eliminating the LTWGS step. In this method, the crude hydrogen stream exiting the HTWGS is passed to a PSA unit. This produces pure hydrogen and captures all of the unwanted materials. The efficiency of the PSA system is dependent on the level of gas impurity, scale of operation the hydrogen purity required and technology specific factors. The quantum of recovered hydrogen is typically quoted in the range 75–85% for a once through PSA system.[12] Table 2.4 has used published data to estimate energy

[12] Sircar S, Golden TC. (2000) Purification of hydrogen by pressure swing adsorption. *Sep Sci Technol* **35**(5): 667–687.

Table 2.4: Estimates for pressure-swing adsorption (PSA) energy efficiencies in producing hydrogen.

	Gas input	Power	Total energy	Steam	H_2	LHV	LHV eff.
	GJ/h	GJ/h	GJ/h	GJ/h	Ncm/h	GJ/h	%
Topsoe compact reformer	15.60	0.43	16.03	0.00	1000	10.22	63.74
Topsoe SMR	12.76		12.76	0.00	1000	10.22	80.11
Lurgi	16.49	0.27	16.76	−0.08	1178	12.03	72.26
	14.00	0.23	14.24	−0.07	1000	10.22	72.26
Technip	12.93		12.93	0.00	1000	10.22	79.04
Foster Wheeler	22.07	0.17	22.24	−4.04	1178	12.03	72.25
	18.74	0.15	18.89	−3.43	1000	10.22	72.25
Foster Wheeler brochure (1)	184.51		184.51	−46.33	11,776	120.34	90.33
	15.67		15.67	−3.93	1000	10.22	90.33
Foster Wheeler brochure (2)	170.84		170.84	−32.70	11,776	120.34	89.58
	14.51		14.51	−2.78	1000	10.22	89.58
NREL data (LHV)	880.35	25.55	905.90	−64.15	66,667	681.28	82.29
	13.21	0.38	13.59	−0.96	1000	10.22	82.29

SMR stands for Steam Methane Reformer, and NREL is the US National Research Energy Laboratory.

efficiencies in PSA units for several proprietary systems for producing hydrogen.[13] This shows energy efficiencies are mainly in the range 70–90%.

Membrane separation systems[14]

In membrane separators, the raw contaminated gas is passed on one side of the membrane and hydrogen selectively diffuses through the membrane to produce a hydrogen-enriched stream on the other side. For polymer membranes, carbon dioxide and water are also usually transported through the membrane so that for hydrogen purification, a series of membrane

[13] *Hydrocarbon Processing — Gas Processing Handbook*, 2004.

[14] Basile A, Nunes SP. eds. (2011) *Advanced Membrane Science and Technology for Sustainable Energy and Environmental Applications*. Woodhead Publishing Limited.

stages are used. This may require inter-stage compression. Membranes are subject to attack by certain impurities in the hydrogen stream and are generally not as robust as PSA.

Membrane systems are cheaper than PSA and have lower operational cost but generally have lower recovery efficiencies. They find their use when only part of a hydrogen stream is required for a downstream process such as the partial removal of hydrogen from a refinery off-gas for recycle.

Some membrane systems are based on palladium metal alloys.[15] These are highly selective for hydrogen, clearly at higher capital cost because of the cost of the palladium alloy.

Cryogenic separation

Should liquid hydrogen (b.p. –253°C) be required then cryogenic separation will remove all of the unwanted materials. Hydrogen can be separated without liquefaction if the gaseous stream is cooled to below the condensation point of all of the other materials present. If helium is present, this will not condense. Cryogenic separation is widely used in separating hydrogen in refinery process streams when the lowest boiling component is typically methane (–165°C) or nitrogen (–196°C).

Hydrogen Production from the Hydrocarbon Process Industry

In Chapter 1, the demand for hydrogen in the refinery and chemical process industries was described. This included the description of the naphtha reformer, which in a refinery is used for the production of high-octane reformate and in the petrochemical industry for the production of aromatics. The naphtha reformer produces large quantities of by-product hydrogen that is sufficient for most of the hydrogen demand in downstream refinery and petrochemical operations. Where necessary, the hydrogen produced within the facilities are supplemented by hydrogen produced from natural gas by methane reforming or partial oxidation. There are other sources of significant volumes of hydrogen in the chemicals industry, which are now outlined.

[15] Grasshoff GJ, Pilkinton CE, Corti CW. (1983) The purification of hydrogen. *Platin Met Rev* **27**(4): 157–169.

Hydrogen Yield from Steam Cracking

The world production of ethylene from steam crackers is in excess of 150 Mt/y. All advanced economies and many developing economies have steam cracker operations. The *Oil and Gas Journal* produces an annual survey of steam crackers that details their operations and feedstock used. The yield of hydrogen differs with feedstock as is shown in Table 2.5.

Table 2.5: Relative hydrogen to ethylene yield for various petrochemical Feedstocks.

Feedstock	Hydrogen/ethylene (w/w)
Ethane	0.132
Propane	0.038
Butane	0.036
Naphtha low severity (LS)	0.043
Naphtha high severity (HS)	0.052
Vacuum gas oil (VGO) LS	0.042
VGO HS	0.042

If ethane is available, it is often a preferred feedstock for the manufacture of commodity polyethylene resins. This feedstock produces the highest yield of hydrogen. The yield of hydrogen from liquefied petroleum gases (LPG) (propane and butane) is lower as is the yield from liquid fuels such as naphtha and vacuum gas oil (VGO)/residua. Liquid fuels have the flexibility of operating in a high severity (HS) or a low severity (LS) mode. This switches the product slate from ethylene to propylene to match market demand for these materials and this also influences the hydrogen yield.

For the most part the world's dominant feedstocks are ethane and naphtha with 30% and 50% respectively of world ethylene production capacity. For a specific location, the preferred feedstock can change overtime. For instance, the United States has seen a great increase in ethane cracking over the past decade as more ethane has become available from shale oil developments.

A typical world-scale steam cracker has a nameplate capacity of 0.5 to 1 Mt/y ethylene so that a 1 Mt/y cracker using ethane would produce 132,000 t/y of by-product hydrogen and a similar scale naphtha cracker about 50,000 t/y of hydrogen.

For stand-alone ethane cracking operations, the by-product hydrogen production exceeds the demand of downstream operations and often the excess hydrogen is downgraded to fuel value. This is useful to the operation by displacing imported fuel oil to fire the cracking furnace. Clearly large stand-alone ethylene crackers could be a significant source of hydrogen for higher value operations.

Hydrogen from Propane and Butane De-Hydrogenation Processes

Butadiene is an important chemical intermediate particularly for the production of synthetic rubbers (e.g. styrene-butadiene-rubber, SBR). Although butadiene is produced as a by-product in the steam cracking of naphtha, this is supplemented by the catalytic cracking of butanes to butenes and then into butadiene. These processes produce co-product hydrogen are illustrated in the following equations:

n-butane to 1-butene[16]

$$+ \quad H_2$$

n-butane to butadiene

$$+ \quad 2H_2$$

[16]The illustration uses the standard stick notation for organic molecules. Only carbon–carbon bonds and hetro-atoms are shown. All other valence bonds are carbon–hydrogen.

There is increasing demand for isobutene as feedstock for producing high octane blending components for gasoline. Isobutene is used to produce alkylate and the ether additive MTBE.

Isobutane to isobutene

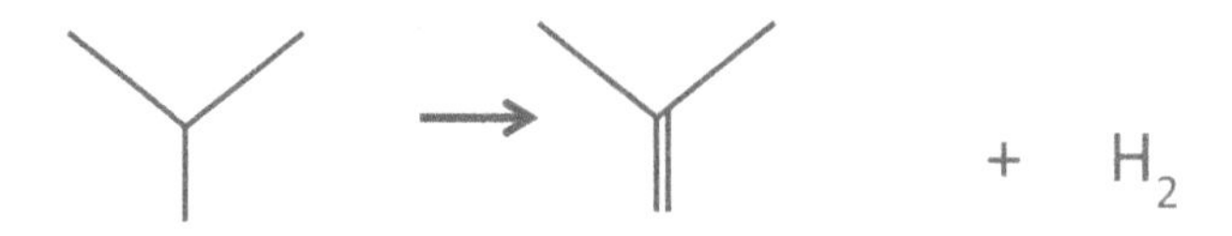

In recent years, in some regions, there has been a move to cracking more ethane than naphtha for the production of commodity polymers. This leaves the petrochemical industry short of propylene for polypropylene and ethylene-propylene copolymers. This shortfall is being increasingly filled by propane de-hydrogenation units to provide the required propylene by catalytic de-hydrogenation.

All these processes are endothermic and require a high operating temperature (typically 500°C). At these temperatures, there is tendency for the catalyst to coke and frequent regeneration is required. There are several technologies available for this. The oldest process (Catofin™ and Catadiene™)[17] uses a chromium catalyst in multiple beds with frequent regeneration. Another process Steam Activated Reforming (STAR) was developed by Philips[18] and uses a tubular reactor system in the presence of steam. A process that has a large and growing use is the Oleflex™ process, which was developed by UOP.[19] In this process, a moving-bed catalytic system is used; Figure 2.13 illustrates the Oleflex process.

In this process, feed (propane or butane) is mixed with some hydrogen to help suppress catalyst fouling and heated to the operating temperature (heater H1) and passed to the first rector (R1). Fresh and regenerated catalyst also enters this reactor and continually moves through typically three or four reactor beds to finally reach a continuous catalyst regeneration unit

[17] Catofin and Catadiene were originally developed from the Houdry catalytic cracking process by Sun Oil. The process is now offered by Clariant: www.clariant.com/en/Solutions/Products/2018/11/28/12/33/CATADIENE

[18] STAR Process by Uhde: www.thyssenkrupp-industrial-solutions.com.

[19] Oleflex was developed by UOP/Honeywell: www.uop.com.

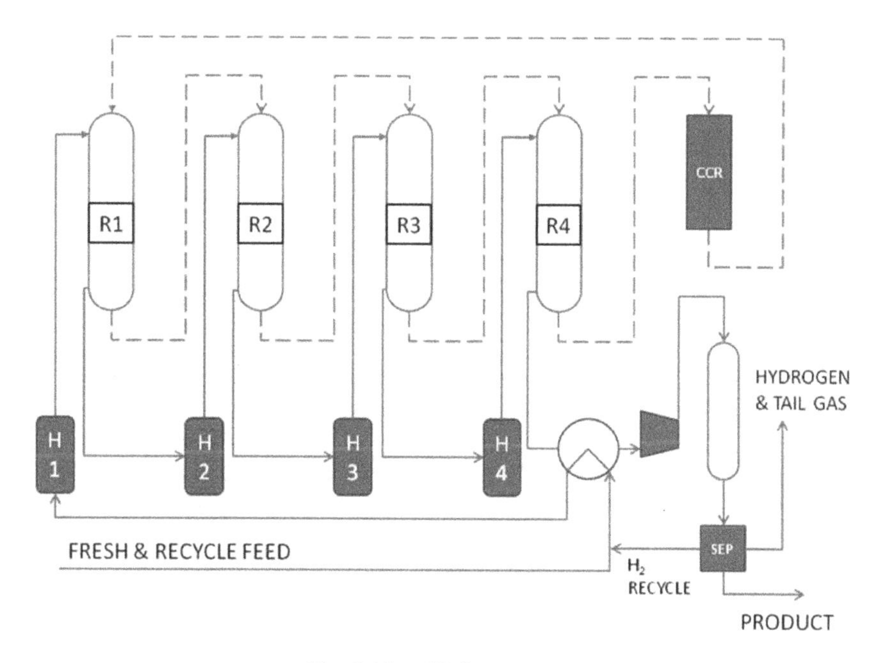

Fig. 2.13: Oleflex process.

(CCR). The endothermic process cools the reactants that are reheated (heaters H2, H3, H4) to the operating temperature. After cooling and compressing the excess hydrogen and cracked gases, they are separated, then product and unconverted feed separated (not shown). The maximum conversion of fresh feed to product is limited by the operating temperature to about 40%.

Hydrogen production

Approximately, 25 Mt/y of olefins are produced by propane and butane dehydrogenation. In theory, this will produce over 1 Mt/y of hydrogen. Since olefins (propylene, butene etc.) are the prime target and since these are easy to transport, the facilities are often remote from other facilities requiring hydrogen. The by-product hydrogen may be downgraded to fuel value and used as a fuel to displace imported fuel oil or gas. There is thus a potential to harness this hydrogen by-product for use in a hydrogen economy.

Carbon Black

Carbon black (tyre black) is produced by the pyrolysis of hydrocarbons. Any hydrocarbon can be used including methane. A favoured feedstock is pyrolysis fuel oil produced as a by-product of steam cracking of naphtha and heavier feeds. In the process part, of the feed is burnt in a restricted amount of oxygen to provide the reaction heat. The process produces by-product hydrogen.

Recently, these processes have been proposed for the co-production of hydrogen and carbon. They typically involve heating methane in a plasma torch to affect the pyrolysis into carbon and hydrogen.[20]

Hydrogen Production from the Chlor-Alkali Industry

Chlorine and sodium hydroxide (caustic soda) are important materials produced by the chemical industry. Chlorine is used to produce bleach, solvents and chloro-polymers such as PVC. Caustic soda is an important chemical in the production of soaps and for refining bauxite to produce alumina. Both chemicals are produced together by the electrolysis of brine. Hydrogen is produced as a by-product. There are three types of electrolytic cell[21]: mercury cells, diaphragm cells and membrane cells.

Because of the issues with mercury in the environment, the older mercury cells are largely being phased out of the industry. In recent times, the use of asbestos in diaphragm cells has discouraged their use and membrane cells are largely replacing both of these technologies. The electrolytic process is illustrated in Figure 2.14.

The cell comprises an anode section and a cathode section separated by a membrane that is porous to sodium ions. This membrane comprises sheets of polymer made from a backbone of poly(perfluoroethylene) with side chains of sulphonic acid or carboxylic acid groups.[22] High purity

[20] Lynum S, Myklebust N, Hox K. *US Patent* 6,068,827 (May 30, 2000) to Kvaerner Engineering AS and also *US Patent* 5,997,837.

[21] Buchner W, Schiebs R, Winter G, Buchel KH. (1989) *Industrial Inorganic Chemistry* (transl. by DR Terrell). VCH.

[22] For example, Nafion™ (du Pont) or Flemion™ (Asahi Glass).

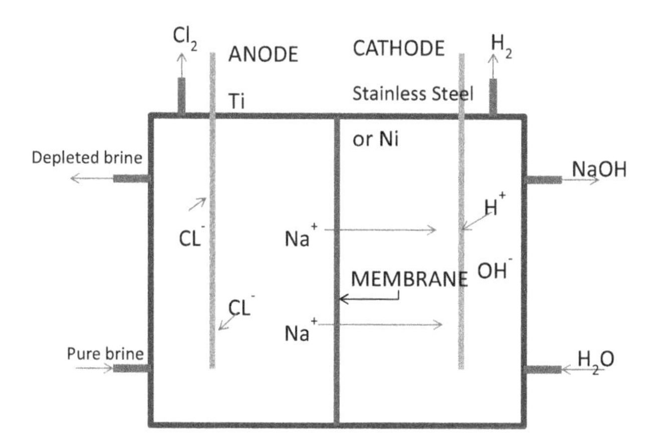

Fig. 2.14: Membrane cell for producing chlorine and sodium hydroxide.

brine[23] enters the cell at the anode side, at the titanium anode, the chloride ions are converted into chlorine gas, which is discharged:

$$2Cl^- = Cl_2 + 2e^-$$

The sodium ions pass through the membrane towards the stainless steel or nickel cathode. Water enters the cathode side of the cell and at the cathode, the water is electrolysed to hydrogen and hydroxide ions.

$$H_2O = H^+ + OH^-$$

The sodium hydroxide solution is withdrawn from the cathode side of the cell.

Absence of the membrane allows mixing of the individual cell components and promotes the formation of sodium chlorate, which also produces by-product hydrogen. The overall reaction being:

$$NaCl + 3H_2O = NaClO_3 + 3H_2$$

Hydrogen Production

The global chorine demand is over 70 Mt/y (2015) and sodium chlorate demand is about 5.5 Mt/y. These values imply about 5 Mt/y of hydrogen by-product is made by the chor-alkali industries.

[23] Membrane cells require brine of significantly higher purity than mercury or diaphragm cells.

Hydrogen in Natural Gas

Perhaps the cheapest and largely untapped source of hydrogen is from natural gas. Hydrogen is rare in natural gas but in some gas provinces hydrogen can be found, sometimes in significant volumes. Why hydrogen is present is a moot point but it can be speculated that it is the result of methane (or higher hydrocarbon) cracking that becomes thermodynamically feasible at high (>800°C) temperatures:

$$\mathbf{CH_4 = 2H_2 + C}$$

The carbon formed in the process remains in the source rock. For this to occur, the source rocks have to be subjected to a suitable high temperature, which may be due to geothermal processes. Sometimes the hydrogen containing natural gas contains significant levels of helium and nitrogen, which suggests that heating to the required temperature may have resulted from radioactivity. Table 2.6 gives an analysis of some gases with a significant hydrogen content.

This data shows that there can be considerable variation in the hydrogen content of some natural gases, with some (the Mount Kitty example) potentially delivering significant volumes of hydrogen.

Table 2.6: Hydrogen in natural gas — compositions from Central Australia.

Gas field	Mount Kitty[24]	Magee-1[25]
Component	Mole%	Mole%
Helium	9	6.24
Hydrogen	11	0.03
Nitrogen	61	43.87
Methane	13	33.49
Ethane	4	6.41
Other hydrocarbons	Not reported	3.41
Total	98	100.00

[24] Santos Limited/Central Petroleum Limited, *Press Release*, 8 May 2014.
[25] Geoscience Australia: data from *Pacific Oil and Gas*.

Observations

The production of hydrogen in the chemical process industries is very large. Large volumes of hydrogen are produced as by-products, much of this is in excess of the demand for downstream operations. In many cases, this excess hydrogen is degraded to fuel value within the facilities producing, it could be available for other uses within a hydrogen-based economy.

CHAPTER 3

THE COST OF HYDROGEN PRODUCTION FROM FOSSIL FUELS

Chapters 1 and 2 outlined the very large requirements for hydrogen in the hydrocarbon processing industries and outlined the methods of production of hydrogen to service this demand from fossil fuels, mainly coal and gas. This chapter addresses the cost of hydrogen production to fulfil the much-expanded demand to run a hydrogen economy. A focus of the chapter will be the cost of production on large-scale facilities to best fit the very large demand for an economy reliant on the use of hydrogen to fulfil transport and energy needs.

A central question for the use of fossil fuels to produce hydrogen is the emission of carbon dioxide to the atmosphere. If this is to be avoided, then the carbon dioxide either has to be used or disposed by burial underground, so called carbon geo-sequestration (CGS). Although there is some interest in the utility of carbon dioxide, few of the proposals would be able to realistically cope with the volumes of carbon dioxide produced. This leaves CGS as the only viable method for disposal. Consequently, a cost estimate for the production of hydrogen without CGS and with CGS will be developed. The approach and the methodology for generating the cost estimates are given in the Appendix.

Hydrogen Production from Natural Gas

The conventional approach to the production of hydrogen from natural gas is illustrated in Figure 3.1.

The figure shows two common approaches to hydrogen production from natural gas based on steam reformer technology. Heat recovery in the reformer generates sufficient process steam for generating power and all other services (e.g. water treatment) required in a utilities module. In fact, in many cases, excess steam is available for sale to third parties or used to produce electricity.

The approach illustrated at the bottom of the figure is the more traditional approach involving a reformer followed by high-temperature and low-temperature water-gas-shift (WGS), carbon dioxide absorption and finally methanation. The more modern approach is illustrated at the top, where after the reformer only HTWGS is used followed by a pressure swing adsorption (PSA) unit producing the hydrogen. PSA off-gas is used as a reformer fuel. Inefficiencies, in the PSA mean that more gas is used for a given quantum of hydrogen but to a first-order approximation, both routes require the same amount of input natural gas for a given quantity of hydrogen.

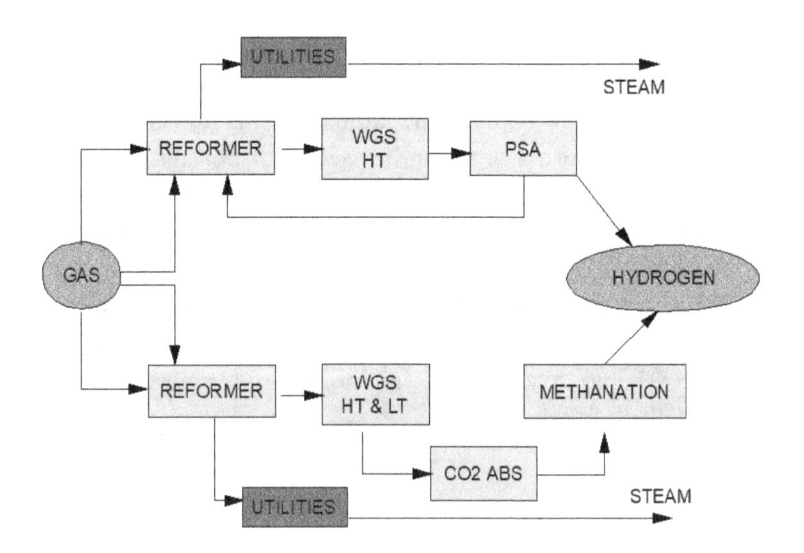

Fig. 3.1: Hydrogen from natural gas.

The Concept of a Central Production Facility

For the most part, the hydrogen economy concerns the use of hydrogen as a transport fuel replacing gasoline and diesel. Many US studies focus on the production of hydrogen on a mass scale for distribution to a large suburban fleet of about 40,000 vehicles from a central facility. Hydrogen need not be produced at the facility, it is just a distribution point. Likewise, the facility itself may service smaller sites for more local distribution. The statistics for the central facility are given in Table 3.1.

To service approximately 40,000 vehicles, about 374 kt/y of hydrogen is required, which is taken as the base case for evaluating various scenarios of hydrogen production. Many workers in the field use different units to express the quantity of hydrogen required for various end-uses. Table 3.2 lists the 374 kt/y in other units.

Table 3.1: Statistics for a central hydrogen production/ distribution facility.

Hydrogen production	t/y	373,940
	t/month	31,161
Vehicles	mpg	65
	km/L	27.47
	L/100km	3.64
	miles/y	12,000
	km/y	19,200
Assume 1 kg H_2 = 1 gal. gasoline	kgH_2/y	184.62
H_2 per vehicle	kg/wk	3.55
Number of vehicle fills		2,025,508
Number of vehicles		38,952

Table 3.2: Central facility hydrogen demand expressed in various units.

Days/year	340
Molecular volume (L/mole)	22.214
kt/year	373.94

(*Continued*)

Table 3.2: (*Continued*)

kt/d	1.100
PJ/y (LHV)	44.84
Mm^3/d (15°C)	13.00
MMcf/d (60°F)	460.1
Nm^3/h	5417745
GJ (LHV)/d	1319018
MW	1526.26

Cost of Hydrogen Production from a Central Facility from Natural Gas

The costs for hydrogen production from natural gas are given in Tables 3.3 and 3.4 for the route employing methanation and the route using a PSA separation system. For ease of analysis, it is assumed that the hydrogen quality from both routes is the same and that trace methane in the methanation route does not influence the efficacy of using this hydrogen in vehicle fuel cells.

The data for the methanation route is taken from data published elsewhere[1] and updated to mid-2018 values. The data for the PSA route is adapted from a 2004 US *National Academy of Sciences* (NAS) report[2] with data brought to 2018 values. Both systems have been scaled to the central facility size.

The methodology used is as that described in the Appendix. The required return on capital (14.6%; see Appendix) is for a three-year construction with a 20-year life and delivering 10% discounted cash flow (DCF). The capital

[1] Seddon D. (2006) The hydrogen economy, hydrogen production from natural gas. *Society of Petroleum Engineers, SPE 101703*, SPE Oil & Gas Conference and Exhibition, Adelaide, Australia 11–13 September.

[2] "The Hydrogen Economy: Opportunities, Costs, Barriers and R&D Needs" National Academies Press (2004).

Table 3.3: Cost estimate for hydrogen production at a central facility.

		Methanation	**Pressure swing adsorption (PSA)**
Hydrogen production	kt/y	373.94	373.94
Construction period	years	3	3
Operating period	years	20	20
Return on Capital (10% DCF) (R)	% Capex	14.60%	14.60%
Location factor	US Gulf	1.0	1.0
CAPEX (Greenfield, 2018)(Cap)	US$M	692.35	532.37
Capital Charges; C = RxCap	US$M/y	101.10	77.74
Working Capital (W)	US$M/y	Nil	Nil
Non Feed OPEX (O)			
Labour (1% Capex/y)	US$M/y	6.92	5.32
Maintenance (3% Capex/y)	US$M/y	20.77	15.97
Insurance (1.5% Capex/y)	US$M/y	10.39	7.99
Catalysts and Chemicals (1% Capex/y)	US$M/y	6.92	5.32
Total OPEX (O)	US$M/y	45.00	34.60
Gas Usage (higher heating value [HHV])	PJ/y	84.76	84.76
Thermal efficiency (HHV basis)	%	62.55%	62.55%
Unit Gas Cost	$/GJ	2.0	2.0
Gas Costs (G)	US$M/y	169.52	169.52
Gross Costs (C + W + O + G)	US$M/y	315.62	281.86
By-product credit (see below)	US$M/y	63.35	63.35
Net Costs	US$M/y	252.27	218.51
UNIT PRODUCTION COSTS	**US$/t**	**674.62**	**584.34**
	$/GJ (lower heating value [LHV])	5.62	4.87

costs are scaled and brought to mid-2018 values from the references given. The methanation route has a higher capital cost than the PSA route due to the higher number of unit operations. This cost estimate may be high since the original data was for a European operating facility rather than the US Gulf, which is used as the reference location. Non feed operating costs (OPEX) are as discussed in the Appendix.

It is assumed that the hydrogen is used immediately by passing-on to another facility and that is there is no storage in the facilities. As a consequence, the working capital for the product inventory is set to zero.

By-Product Credits

Both these routes operate at about 63% thermal efficiency,[3] which means 37% of the energy in the gas is lost in the process. Much of this is captured by steam raising in the reformer and the HTWGS. This steam is used to supply the facility demand for process steam and to generate electricity for pumps, lighting etc. There is excess steam that generates the potential for a by-product credit. In determining the viability of projects in the process industries, the optimisation of by-product credits is often critical. Excess steam can be sold 'over-the-fence' to other unit operations or to third parties or converted into electricity at notional grid prices this is the method adopted here. High temperature, relatively high-pressure process steam can be converted into electricity with a thermal efficiency of about 25% whilst lower temperature, lower pressure process steam has a much lower efficiency of conversion. Clearly, the optimisation of a process steam system is complex and beyond the scope of this work hence the result here will be subject to some error. The aim is to give an order of magnitude estimate for the value of the excess steam produced. The estimate is given in Table 3.4 and with an electricity value of 5c/kWh ($50/MWh), this generates a by-product credit of US$63.35M.

[3]A value for the thermal efficiency in the low 60% range is typical for greenfield operations for this type of facility. Quoted higher efficiencies are usually for the Inside Battery Limits (ISBL) plant and do not take into account utilities and off-sites or use an imported utility such as power.

Table 3.4: Estimated by-product steam, power generation and value for base case.

Reformer excess steam	PJ/y	17.20
Electricity equivalence (25%)	PJ/y	4.30
	kWh/y	1.19E + 9
Electricity value	cents/kWh	5
Sub total	$M/y	59.73
Steam from water-gas-shift (WGS)	MWh	75,544
Value as power (10% efficiency)	$M/y	3.62
Total	$M/y	**63.35**

This by-product credit considerably reduces the net production cost of the hydrogen.[4]

No other by-product or credit is considered for this base case. However, in practice, other credits may arise from

- Possible sales of synthesis gas to third parties in developed locations.
- Sale of natural gas liquids that may be extracted from the input gas.
- Sale of sulphur that might be present in the input gas that requires extraction prior to the reformer.
- Sale of extracted carbon dioxide to a downstream user; e.g. for enhanced oil recovery (EOR).

Gas Price

Note that the natural gas is priced on a higher heating value (HHV) basis, whilst the hydrogen product is priced on a lower heating value (LHV) basis to better represent its value as a vehicle fuel.

For the base case, gas is priced at $2/GJ (HHV), which is similar to prices for world-scale Liquefied Natural Gas (LNG) export projects and similar to current Henry Hub prices at the time of writing (late 2020).

[4] Ward R, Seers N. (June 2002) Hydrogen plants for the new millennium. In: *Hydrocarbon Engineering*, gives a significant insight into the factors influencing steam export from steam reformers.

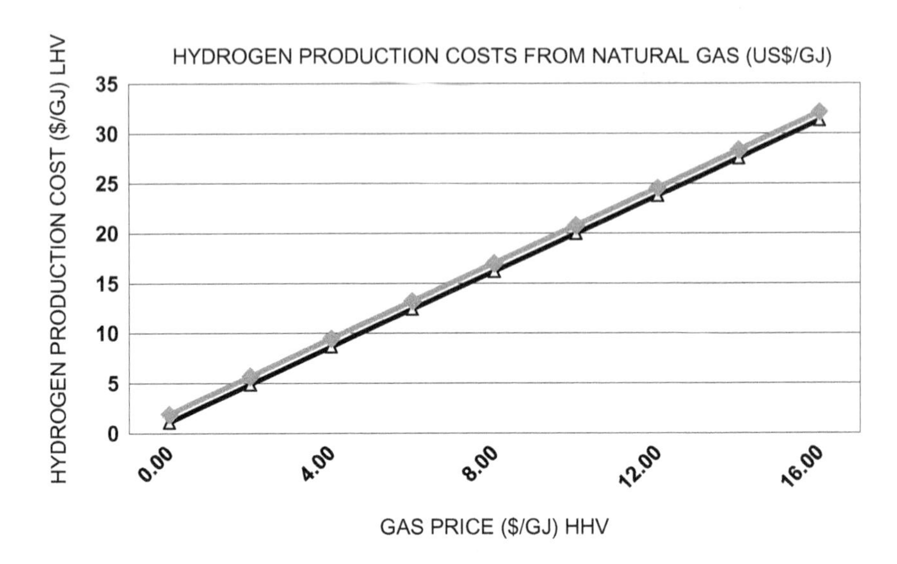

Fig. 3.2: Sensitivity of hydrogen production cost to gas price.

With this price, the unit production cost of hydrogen is estimated at $675/t ($5.62/GJ: LHV basis) for the methanation route and $584/t ($4.87/GJ) for the PSA route. Without the steam, by-product credit the estimated production costs are $844/t ($7.04/GJ) and $754/t ($6.28/GJ), respectively. This may be the situation in a remote location where it is impossible to realise this credit.

The production cost of hydrogen is very sensitive to the gas price. The fixed variable relationship for the base case is illustrated in Figure 3.2. Both methanation and PSA are very similar; the slight difference between the two lines results from the slightly higher capital cost for the methanation route.

In many parts of the world, gas is unavailable at the low $2/GJ (HHV) base case, which generates hydrogen production costs in the region of $5/GJ. For many areas, the cost of gas is $8/GJ or higher. In these circumstances, hydrogen production cost lies between $16 and $17/GJ (LHV).

Capital Cost

The other main variable in the cost of hydrogen production is the capital cost. Much R&D effort has gone into making process improvements and whist minor changes and improvements do occur as more facilities are built (a learning curve, e.g. by improvements to the pass-efficiency of PSA units), it is unlikely there would be a dramatic technology breakthrough in this field. The quoted 2004 NAS study quoted modelled a hydrogen process that envisaged a 20% lowering of the capital cost but this does not seem to have been achieved by 2020.

Greenhouse Emissions from Steam Methane Reforming (SMR)

The main issue for producing hydrogen for a Hydrogen Economy from fossil fuels is minimising and preferably eliminating carbon dioxide emissions. This issue of greenhouse gas (GHG) mitigation and its cost is discussed more fully later in this chapter. For now, if we accept that the elimination of GHG emissions is desirable, then this will influence the approach to producing hydrogen from fossil fuels. This is illustrated by the cases as outlined in Figure 3.1. Inspection of the two alternative routes indicates that although both routes have a reformer that requires fuel combustion, this produces an exhaust stream (flue gas) from which it is difficult and costly to remove the carbon dioxide. The methanation route captures a pure carbon dioxide stream, from the carbon dioxide stripper, which could be more easily geo-sequestrated. Thus, the higher capital cost of the methanation route could have the advantage of lower atmospheric emissions of carbon dioxide.

The design could go further by opting to use a partial oxidation unit rather than a reformer.[5] This will use the same quantity of gas but all of the carbon input would be captured in a carbon dioxide stripping unit and be more easily geo-sequestrated; Figure 3.3.

Taking into account the need to fire a steam reformer, this partial oxidation route uses very similar amounts of gas to produce the same amount of hydrogen. The capital cost of the partial oxidation gasifier (POX), its associate waste-heat boiler (WHB), the air-separation unit (ASU) and WGS is very similar to the cost of a steam reformer with HT and LT WGS

[5] See Shell Blue Hydrogen Process: https://www.shell.com/business-customers/catalysts-technologies/resources-library/q-n-a-affordable-blue-hydrogen-production-with-shell-blue-hydrogen-process.html.

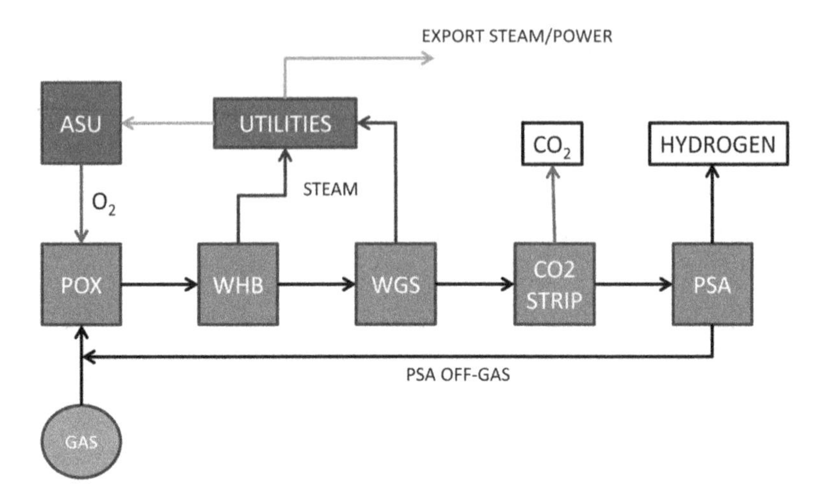

Fig. 3.3: Hydrogen from natural gas incorporating carbon dioxide capture.

so that the only additional cost is for the larger carbon dioxide stripping unit. Most of the off-gas can be recycled to the gasifier with only a small volume of the off-gas extracted to prevent the build-up of inert materials in the recycle loop.

Overall, it would be expected that this route would have very similar economics to the steam reformer system using methanation and offering carbon dioxide capture for most of the carbon input (possibly >95%). Note, however, that many case studies using POX (or Auto-thermal reforming) fail to take account of the capital cost of the ASU; the oxygen being imported from a third party.

Impact of Scale of Operation

The above base case modelling is for a large plant supporting hydrogen production for a large central distribution facility. There is some interest in producing hydrogen on a smaller scale supporting fewer vehicle filling operations. Going to a smaller-scale operation will increase the unit production costs due to (i) loss of the economy of scale factor and (ii) the generally higher gas costs for smaller volume off-take. Figure 3.4 illustrates the effect of reducing the scale. The figure plots the estimated

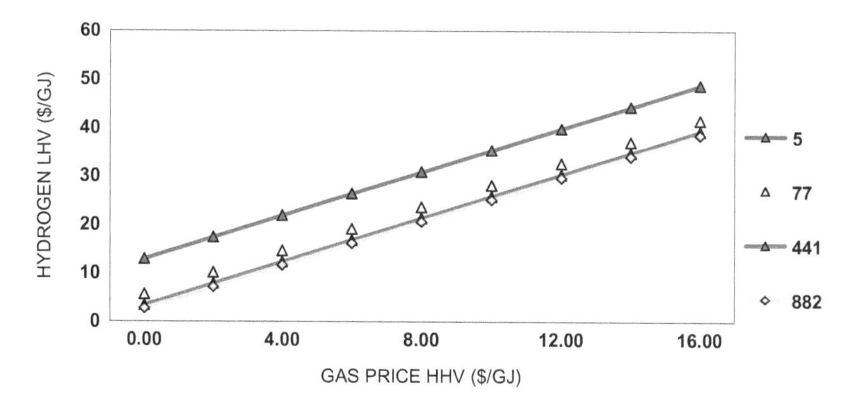

Fig. 3.4: Influence of scale and gas price on hydrogen production cost.

hydrogen production cost against gas price for various scales ranging from 5.2 t/d (0.21 PJ/y), 77 t/d (3.14 PJ/y), 441 t/d (18 PJ/y) and 882 t/d (36 PJ/d) of hydrogen production using a steam reformer/methanation route. The scaling factor used was 0.786.

The largest scale 882 t/d is 300 kt/y and is very similar to the base case for a central facility. As can be seen from the graph the economics rapidly deteriorate at scales below 77 t/d (3.14 PJ/y). For large-scale operations, gas is generally available at $4–$6/GJ (HHV), which delivers hydrogen in the range $10–$14/GJ (LHV). For smaller-scale operation, gas prices would be typically $8/GJ (HHV) or higher, which means that hydrogen production cost is estimated at over $20/GJ (LHV).[6] Carbon capture is not considered viable for the smaller-scale operations.

Hydrogen from Coal

The production of hydrogen from coal in the OECD has been largely displaced by natural gas as a feedstock. The main use of hydrogen from coal plants is in the production of ammonia. The technology used is commonly a moving bed gasifier driven by air. This introduces nitrogen into the

[6] Note gas is invariably sold on a HHV basis whereas for hydrogen-fueled vehicles, we are concerned the LHV production cost.

synthesis gas stream. For many decades, in The Republic of South Africa coal has been converted into transport fuels using the Fischer-Tropsch process. In China, in the past decade, several coal to olefins facilities have been built. The first step of these processes is oxygen driven coal gasification to synthesis gas, which is followed by WGS. This is the starting point for the production of hydrogen from coal.

The main issue for coal feedstock is that for a given quantum of hydrogen, considerably more capital is required compared to natural gas facilities. This leads to consideration of very large-scale operations, which, in theory, would match the volume demands of a hydrogen economy better than alternative approaches using natural gas. Furthermore, coal is more ubiquitous and generally cheaper than alternative feedstocks. Unfortunately, far more carbon dioxide is emitted from coal-based plants, this would have to be removed in a carbon constrained world.

Cost of Hydrogen Production from Coal

The Parsons company for the US Department of Energy, National Energy Technology Laboratory has produced a major study on the use of various gasifier types and various coals for the production of hydrogen.[7] The study gives process flows, equipment sizes and cost analysis. For this book, this level of detail has been simplified.

The production of hydrogen from coal involves more unit operations than hydrogen manufacture from natural gas. Figure 3.5 illustrates the method modelled.

Coal and oxygen (from an ASU) are fed to a gasifier and the raw gas cleaned up by an appropriate method (see Chapter 2). This is followed by a high-temperature WGS unit, which optimises the hydrogen production. An acid-gas plant removes hydrogen sulphide and carbon dioxide. The hydrogen sulphide is passed to a Claus plant (not shown), which produces sulphur for export. The carbon dioxide is obtained as a pure stream and can be compressed for geo-sequestration. The hydrogen-rich gases are

[7] Longanbach JR, Rutowski MD, Buchanan TL, Klett MG, Schoff RL. Capital and operating costs of hydrogen production from coal gasification. Final Report April 2003, DOE Task 50611.

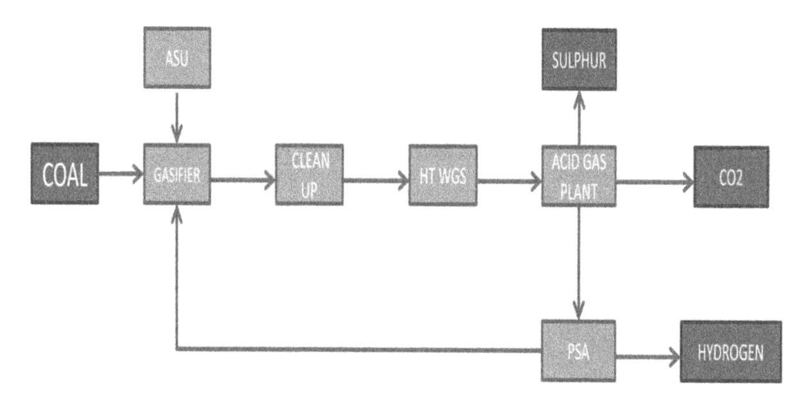

Fig. 3.5: Hydrogen from coal.

passed to a PSA unit that extracts the hydrogen. The tail gas is passed back to the gasifier to convert any methane present and recover any hydrogen passing through the PSA.

There are many variants on this flow-sheet. For instance, the use of a cold methanol wash in the clean-up. This can remove carbon dioxide and hydrogen sulphide. The hydrogen sulphide and carbon dioxide can be combined and both geo-sequestrated. The PSA system can be replaced with a low-temperature WGS followed by a carbon dioxide stripper unit and methanation reactor. The comparison of three studies is given in Table 3.5.

The first two results are from published Shell work and the third from an *NAS* paper.[8] The Shell study used the same gasifier size (Shell entrained-bed) gasifier and modelled two options, the first with a PSA unit and the second employing a methanation system. The first option produced significantly more hydrogen than the second option (1042 kt/y vs 835 kt/y).

Note these values are double the size for the base case natural gas. The NAS study was on a smaller scale at 408 kt/y, which is on a par with the base case natural gas study given above. The estimate follows the method outlined in the Appendix with the all the data brought to mid-2018 costs. Note coal plants take longer to build than gas plants so the required capital

[8] "The Hydrogen Economy Opportunities, Costs, Barriers and R&D Needs", National Academy of Sciences, 2004.

Table 3.5: Economics of hydrogen production from coal without carbon capture and storage (CCS).

Source		**Shell**	**Shell**	**National academy of sciences (NAS)**
		PSA	**Methanator**	**Pressure swing adsorption (PSA)**
H_2 production nominal capacity	kt/y	1042.72	838.90	408.00
	kt/d	3.0668	2.4673	1.2000
	PJ/y (lower heating value [LHV])	125.08	100.63	48.94
	MMcm/d	30.46	24.50	11.92
	MMcf/d	1077.62	866.97	421.65
	Ncm/h	1,268,978	1,020,925	496,530
	GJ/d (LHV)	367,877.2	295,966.5	143,944.3
Construction period	Y	4	4	4
Required return	%	10%	10%	10%
Operating period	Y	20	20	20
Capital recovery	%	15.15%	15.15%	15.15%
FINAL CAPEX	$M	3,321.24	3,321.24	1,632.00
Capital Charges	$M/y	503.13	503.13	247.23
Working Capital (% Capex)	$M	0	0	0
CAPITAL CHARGES (C)		503.13	503.13	247.23
Operating Costs	% Capex			
Labour	2.51%	83.03	83.03	40.80
Maintenance	3.5%	116.24	116.24	57.12
'Insurance'	1.5%	49.82	49.82	24.48
Catalysts and Chemicals	1%	33.21	33.21	16.32

Table 3.5: (*Continued*)

Source		**Shell**	**Shell**	**National academy of sciences (NAS)**
		PSA	**Methanator**	**Pressure swing adsorption (PSA)**
NON FEED OPEX (O)		282.31	282.31	138.72
Non Gas Feedstocks				
Electricity imports	MWh/y	1,241,015	1,241,015	198,300
Non Gas Feedstock Costs	\$M/y	62.05	62.05	9.91
Other	\$M/y	9.89	9.89	0.00
TOTAL NON FEED COSTS	\$M/y	847.49	847.49	395.86
COAL USAGE	PJ/y (LHV)	244.79	244.79	95.39
Thermal Efficiency (%)	%	60.40%	48.59%	60.64%
COAL PRICE	\$/GJ	1.00	1.00	1.00
COAL COSTS	\$M/y	244.79	244.79	95.39
UNIT PRODUCTION COST CALCULATION				
Gross Costs	\$M/y	1092.27	1092.27	491.26
By-product Credits (B)	\$M/y	3.02	3.02	5.34
Net Costs	\$M/y	1089.26	1089.26	485.92
H_2 UNIT PRODUCTION COST	\$/t	1044.63	1298.44	1190.98
	\$/GJ	8.71	10.82	9.93

return is higher. The hydrogen is used immediately so the working capital is set at zero. Labour and maintenance charges are set to reflect the higher operating costs of coal plants. All three cases use imported power at a unit cost of \$50/MWh.

The thermal efficiencies of coal operations are lower than natural gas and these are estimated at about 60% for the first Shell case and the

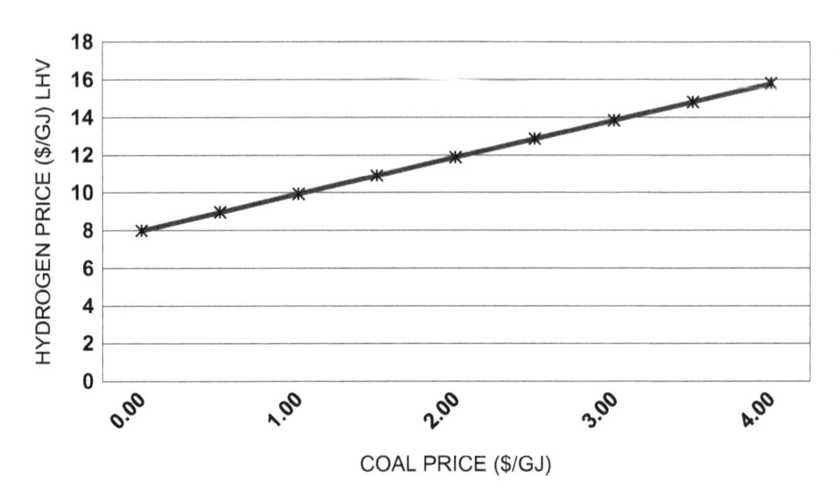

Fig. 3.6: Hydrogen from coal — sensitivity to coal price.

NAS study. These values are high as the import power has not been included in the estimate.

Sulphur produced as a by-product generates a by-product credit. The sulphur is valued at $100/t. The NAS value is higher because of the higher sulphur content coal used in that study. The estimated hydrogen production costs are in the range $1000/t–$1300/t ($8.7/GJ–$10.8/GJ).

Sensitivity to Coal Price

The sensitivity to the coal price is illustrated in Figure 3.6 for the NAS study. For coal at $1/GJ, which is approximately $25/t, and typical for coal mined (open-cut) juxtaposed to a hydrogen production plant,[9] the hydrogen, production cost is estimated to be about $10/GJ. Export coal varies between $50/t and $100/t, which is about $2–$4/GJ. At these prices, the hydrogen production cost lies between $12/GJ and $16/GJ.

Greenhouse Gas Emissions

The current principal driver for a hydrogen economy is the elimination of GHG emissions. This driver replaces the earlier concern of rising oil

[9]This is typical for open-cut operations feeding directly into coal generation facilities.

prices as a result of actions by OPEC in the 1970s and early 1980s by finding alternatives to gasoline and diesel fuels. The Kyoto (1995) and Paris (2015) Protocols aim to limit climate warming by GHGs to 2°C by restricting the emission of gases with a high warming potential. The production of hydrogen from fossil fuels converts all of the carbon present in the fuel into carbon dioxide and this represents the major hurdle to use fossil fuels as the hydrogen source, unless it can be disposed of in some manner for at least 100 years.

There are several gases with high greenhouse warming potential (GWP) and of principal concern to hydrogen production is carbon dioxide, methane and to a lesser extent nitrous oxide.

Carbon dioxide is the principal GHG of concern and other gases of greenhouse potential are marked relative to it on a mass basis. Carbon dioxide is the result of burning fossil fuels in furnace operations and similar (steam reformers) or in the conversion of synthesis gas into hydrogen by the WGS reaction. Many natural gas resources contain carbon dioxide, often in very high amounts, and upon development of a gas field the carbon dioxide is extracted and discharged to atmosphere. This additional carbon dioxide emission has to be taken into consideration if natural gas is used to make hydrogen.

Methane has 28 times the GWP of carbon dioxide and is the principal component of natural gas.[10] Methane emissions to the atmosphere occur in natural gas recovery as fugitive emissions from gas well development and well operation, gas treatment and pipelines. Methane is often associated with coal beds and is emitted to atmosphere by coal mining. Generally, the deeper the coal measure the more methane is present in the coal seam. This can be recovered as coal seam gas (CSG) and is often drained ahead of underground mining operations as a safety measure. Open-cut mines also emit methane to atmosphere.[11] All these fugitive

[10]The GWP values are inexact. There are several metrics for determining the relative greenhouse potential of gases. The time horizon is an important factor. For 20 years, the GWP of methane is 84. Here a 100-year time horizon is used. '*Climate Change 2013. The Physical Science Basis,*' IPCC 2013, contribution to Working Group 1 details the IPCC method.

[11]Saghafi A. (2013) *Estimation of fugitive emissions from open cut coal mining and measurable gas content.* 13th Coal Operators' Conference, University of Wollongong,

emissions have to be considered in any assessment of the relative merits of emissions between alternative strategies for producing hydrogen.

Nitrous oxide with a GWP of 265 times that of carbon dioxide is a minor component of the mixed nitrogen oxides (NOx) produced on burning fuel (any fuel) in air. NOx is the result of high-temperature combination of nitrogen and oxygen in air and the quantity produced is dependent on the flame temperature and burner type.[12] Nitrogen oxide emissions from natural gas fired furnaces are relatively high because methane burns at a higher temperature than other fossil fuels. Note that hydrogen burns at a very high temperature so it would be expected that burning hydrogen in air would produce larger amounts of NOx.

With the exception of the recovery of natural gas,[13] the major portion of greenhouse emissions in hydrogen production result in the burning of fossil fuel and occurs as carbon dioxide and this is the focus of the following account.

Greenhouse Emissions in the Production of Hydrogen

In the conventional production of hydrogen, GHG emissions occur as relatively pure streams of more than 90% carbon dioxide and dilute, impure streams of less than 20% carbon dioxide. The dilute streams contain various other gases, principally nitrogen. These streams are treated in different ways to produce concentrated liquid carbon dioxide for disposal.

Carbon dioxide capture from dilute, impure streams

Dilute, impure steams result from fuel-air combustion in furnaces with the duty to heat process streams or from boilers used to generate process

---

The Australasian Institute of Mining and Metallurgy & Mine Managers Association of Australia, 306–313.

[12] Tacas TJ, Juedes DL, Crane ID. (June 21, 2004) *Oil & Gas Journal*, p. 48.

[13] Some jurisdictions allow the emission of methane to atmosphere in well testing activities and many permit open combustion of well gases during testing and development. However, on many gas fields, it is a common practice to extract carbon dioxide present and emit the gas to atmosphere.

steam and/or power. Air combustion also introduces NOx into the flue gases. Incomplete combustion, especially for natural gas, can leave methane in the flue gas. Especially if using fuel oil, sulphur oxides (SOx) are introduced into the flue gases. The usual practice is to operate furnaces with a slight excess of air relative to that required for ideal combustion. This leads to maximum efficiency and reduces particulate emission (smoke) from the flue stack. However, this practice introduces a small amount of oxygen into the flue gases. Another aspect of air combustion is that the flue gas containing the carbon dioxide requiring capture is diluted with a major excess of nitrogen. This greatly increases the size of the plant required to treat the flue gas and greatly increases the capital cost.

By way of example, a natural gas burner using a gas of composition methane 90% (vol) and carbon dioxide (10%) operating with 10% excess air would produce a fuel and flue gas with compositions illustrated in Figure 3.7.

The aim of carbon capture is to remove as much of the carbon dioxide in the flue gas as possible. As illustrated in the figure, the carbon dioxide is a minor component of the flue gas; it comprises only 9.2% of the flue gases. Prior to carbon dioxide absorption it would be necessary to remove as much of the water as possible as this would dilute the solvent used in the most common absorption systems that would increase the energy

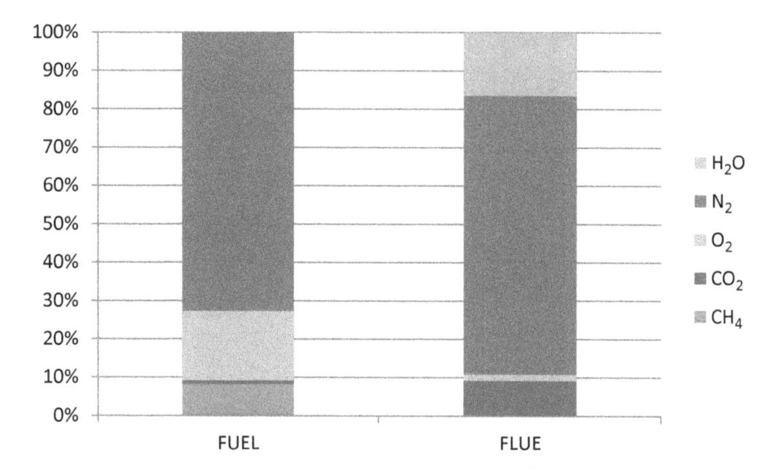

Fig. 3.7: Fuel and flue gas compositions for a 90% methane, 10% carbon dioxide fuel gas operating with 10% excess air.

demand in the solvent regeneration unit. This would require cooling the large volume of flue gas to condense the water. The carbon dioxide then has to be removed from a large excess of nitrogen (72.7%), which contains residual oxygen (1.7%). Along with NOx and SOx, this residual oxygen attacks most the common solvents producing a range of compounds that causes foaming, fouling, increases in solvent viscosity and promotes corrosion in a carbon dioxide stripper.[14]

This example uses a flue gas with 9.2% carbon dioxide. This is typical for coal boilers where carbon dioxide levels can range from about 9% to 15% by volume. However, this is very high compared to gas-turbine generation where the necessity to have a large air by-pass to cool the system results in carbon dioxide contents of 3% or less.

A process that compares the use of the common solvent (methyl ethanolamine, MEA) with one under development and may be considered as an optimum, is illustrated in Figure 3.8[15] with data presented in Table 3.6.

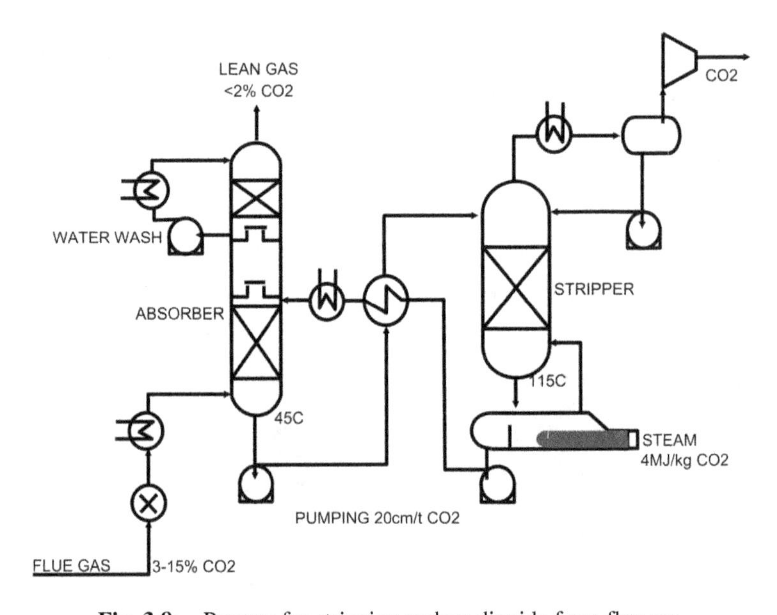

Fig. 3.8: Process for stripping carbon dioxide from flue gas.

[14] Strazisar B. R, Anderson R.R, White C.M. (2003) *Energy and Fuels* **17**: 1034.
[15] Saxena MN, Flintoff W. (December 2006) *Hydrocarbon Processing*, p. 57.

Table 3.6: Data for amine treatment of flue gases to remove carbon dioxide.

	Methyl ethanolamine (MEA)	Special amine
Steam for solvent (t/tCO_2)	1.95–3.0	1.2
Regeneration fuel (GJ/tCO_2)	4.2–6.5	2.6
Solvent flow (m^3/tCO_2)	17–25	11
Power for pumps (kWh/tCO_2)	150–300	198
Cooling water (m^3/tCO_2)	75–165	150
Solvent consumptions (kg/tCO_2)	0.45–2.0	0.35
SO_2 tolerance (ppm)	10–100	<10

In this process, which is based on standard amine (MEA) absorption technology, flue gas enters an absorber where the carbon dioxide is stripped out of the flue gas to a concentration of less than 2% volume. A water wash in the top of the column removes and recycles amine to the absorber. The carbon dioxide rich solution is passed to a stripper boils the solvent expelling the carbon dioxide. Excess water is also removed by this process. The carbon dioxide is then compressed for possible geo-sequestration.

Detailed economic comparison of different absorber systems applied to flue gas system in Venezuela has been described by Aldana *et al.*[16] The objective was to produce carbon dioxide for an EOR scheme, three approaches were studied. The first two used an inhibited MEA system and the third added a deoxy unit to remove residual oxygen. The flue gas source was cement kilns. This work has been brought to 2018 values and to a US Gulf location and detailed in Table 3.7. This gives estimates for the cost of extracting and delivering liquid carbon dioxide between $86/t and $102/t.

One of the surprising features of these studies is very high estimates for the cooling water requirement (not explicitly shown but contributing to the high utility charges). This is a consequence of having to reduce the

[16] Aldana G, Arai B, Elliot DG. (1985). An Evaluation of CO_2 for EOR in Venezuela, In: SA Newman (ed.), *Acid and Sour Gas Treating Processes.* Gulf Publishing Company, p. 164.

Table 3.7: Estimates for carbon dioxide extraction cost from flue gas.

		DOW/ methyl ethanolamine (MEA)	**UCC/ MEA**	**DEOX/MEA**
Carbon Dioxide	Mt/y	1.79	1.79	1.79
Capex (2018)	$M	$253.82	$248.64	$226.14
Return on Capital (ROC)	$M/y	$35.12	$34.41	$31.29
OPEX	$M/y	$58.55	$51.58	$45.42
Maintenance	$M/y	$7.61	$11.65	$10.60
Utilities	$M/y	$80.84	$71.72	$66.76
TOTAL COSTS	$M/y	$182.13	$169.36	$154.07
	$/t CO_2	$101.86	$94.71	$86.16

flue gas temperature to near ambient to remove a major portion of the water present in the flue gas prior to the carbon dioxide removal unit.

There have been many approaches proposed to date for the removal of carbon dioxide from flue gas, and many proposed solutions, especially for stripping carbon dioxide out the flue gases of large thermal power plants. None so far none have proved satisfactory due to a combination of poor efficiency and high capital and operating cost.[17]

Carbon dioxide capture from concentrated, relatively pure streams

If hydrogen is to be made from natural gas, concentrated steams are found in gas field production facilities where acid-gas plants have the duty to remove carbon dioxide and hydrogen sulphide that might be present in the gas. This is necessary prior to natural gas transmission by pipeline in order to avoid pipeline corrosion. Typical pipeline specifications are for the carried gas to be less than 3% (vol.) carbon dioxide and hydrogen sulphide less than 10 mg/m^3. Mostly the carbon dioxide is discharged to

[17] Ogawa T, Ohashi Y, Yamanaka S, Miyaike K. (2009) *Energy Procedia* **1**: 721; Songolzadeh M, Soleimani M, Ravanchi MT, Songolzedah R. (February 2014) *The World Scientific Journal.* Development of Carbon dioxide removal system from the flue gas of coal fired power plant.

atmosphere and the hydrogen sulphide converted into sulphur by the Claus process. In some instances where this would result in very large discharge of carbon dioxide, the carbon dioxide is compressed and liquefied and injected into geological strata beneath the producing gas field.[18]

Concentrated streams also occur in the conversion of synthesis gas to hydrogen as a consequence of WGS. This reaction converts carbon monoxide and water (steam) into carbon dioxide and hydrogen. The carbon dioxide is removed in a downstream scrubber unit, which is very similar to those used in a natural gas acid-gas plant, and which produces carbon dioxide as a concentrated stream. The carbon dioxide can easily be captured and compressed for underground or similar disposal.

Because the process streams contain less impurities than flue gas, there are more technical approaches to the design of carbon dioxide removal systems in order to produce a carbon dioxide stream suitable for compression and disposal. Many are similar to hydrogen purification systems and some of the more common approaches are described.

Membrane Systems

Membrane systems are relatively cheap and compact and find utility in space limited situations such as offshore gas production and as a pretreatment to a main separator system. Membrane systems are best suited for streams with carbon dioxide content over 15% (vol.) and at 700 kPa or higher pressure. The system is illustrated in Figure 3.9.

The unit comprises two sections separated by a membrane that will selectively allow the passage of carbon dioxide to pass from the high-pressure inlet side to the lower pressure side. With the common cellulose membranes, some other gases also pass through the membrane such as hydrogen and water, which limits its applicability to hydrogen purification.[19] The result is that the low-pressure side has a gas that is enriched in carbon dioxide and high-pressure side becomes depleted. The system is often used with several membranes in series, with intermediate compression as may be needed.

---

[18] Carbon dioxide is injected as a dense phase rather than as liquid. For example, Sleipner gas field offshore Norway and Gorgon gas field offshore NW Australia.

[19] For hydrogen purification, hydrogen selective membranes are used.

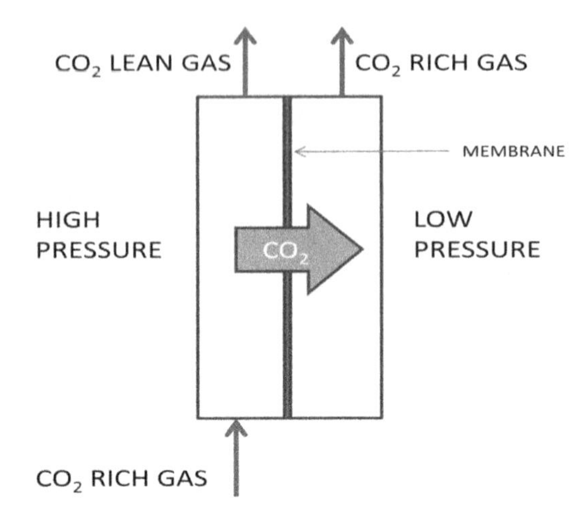

Fig. 3.9: Membrane system for removing carbon dioxide.

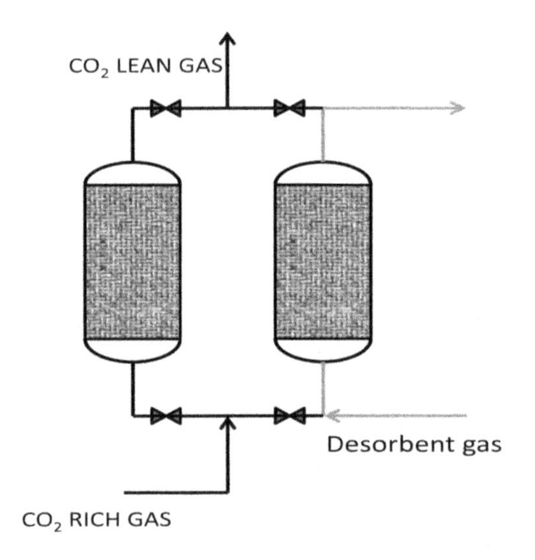

Fig. 3.10: Adsorption system for carbon dioxide removal.

Adsorption Systems

Adsorption systems are similar to the PSA systems described for hydrogen purification. They are best used for streams with carbon dioxide content over 30% and at pressures over 500 kPa. The system is illustrated in Figure 3.10.

A carbon dioxide rich gas is passed to an adsorber where the solid adsorbent removes the carbon dioxide. When the absorbent is saturated, the stream is switched to another absorber and the saturated absorbent is regenerated by driving off the adsorbed carbon dioxide; e.g. by using steam.

Absorption Systems

Absorption systems are the most common in the production of hydrogen. In the process, the gas is contacted with a liquid that dissolves the carbon dioxide. The solvent is regenerated and the carbon dioxide is expelled. The basic system is illustrated in Figure 3.11.

There are many variants on this process but there are two general approaches. The first is physical absorption in which the gases dissolve in the absorbing fluid according to their relative boiling point. Hydrogen and other gases of low boiling point such as nitrogen are not well absorbed whereas gases of higher boiling point such as hydrocarbons, carbon dioxide[20] and hydrogen sulphide more readily dissolve in the liquor. Physical systems are generally applied to gases where the carbon dioxide is over 20% and at pressures over 200 kPa.

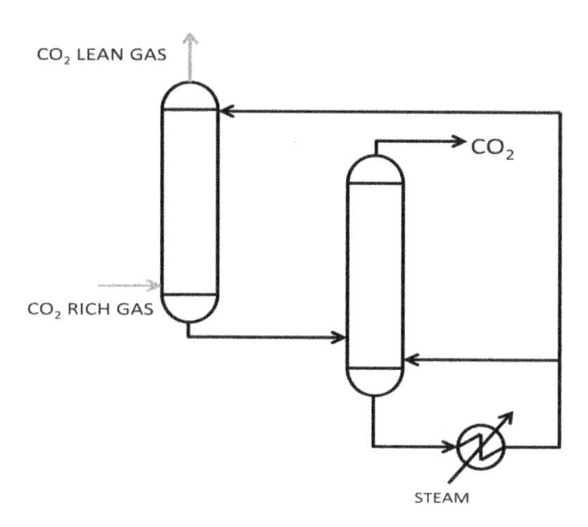

Fig. 3.11: Basic system for absorption process for carbon dioxide removal.

[20] For these systems, carbon dioxide is considered to have a boiling point of 175K.

There are several solvents in common use such as polyethylene glycol derivatives (DEPG), propylene carbonate (PC), n-methyl pyrrolidone (NMP) and methanol. The use of chilled methanol (typically about –40°C) is common in cleaning raw synthesis gas coming from coal gasifiers (Rectisol ™ process) as it readily removes higher boiling liquors and tars from the raw gas.

The other approach is to recognise that the two gases of concern (carbon dioxide and hydrogen sulphide) are acidic and that alkaline sorbents will preferentially extract these materials. This enhanced absorption capacity facilitates its application to streams of lower pressure and lower carbon dioxide content. A common solvent is an aqueous solution of MEA and derivatives and is widely used in natural gas treatment plants. Another based on a solution of potassium carbonate is widely used in removing carbon dioxide from ammonia synthesis gas streams (Benfield process).

Depending on the design, the carbon dioxide and hydrogen sulphide can be produced as a pure streams or a mixed steam of the two gases.

Cryogenic Processes

Carbon dioxide can be separated and purified using cryogenic processes at extremely low temperature. In the process, the gas is compressed and expanded to cool the gases to the point of carbon dioxide liquefaction. Liquid carbon dioxide can be passed to a distillation column that produces a pure carbon dioxide product. Gases with low boiling points such as hydrogen, nitrogen and methane are not condensed. Impurities of higher boiling point dissolve in the liquid carbon dioxide. The cryogenic process is claimed to have a low specific energy consumption when compared with absorption systems.[21]

Cost of carbon dioxide removal from hydrogen process streams

Aldana *et al.*[22] have described the removal cost of carbon dioxide from hydrogen off-gas for EOR operations in Venezuela using a variety of

[21] Xu G, Liang F, Yang Y, Hu Y, Zhang K, Liu W. (2014) An improved CO_2 separation and purification system based on cryogenic separation and distillation theory. *Energies* **7**: 3484.
[22] Aldana G, Arai B, Elliot DG. (1985) An Evaluation of CO_2 for EOR in Venezuela. In: SA Newman (ed.), *Acid and Sour Gas Treating Processes*. Gulf Publishing Company, p. 164.

Table 3.8: Cost estimates for extraction of carbon dioxide from hydrogen off-gas.

		Diethanolamine (DEA)	Benfield	Selexol	Triethanolamine (TEA)	Membrane	Cryogenic
CO_2	Mt/y	1.681	1.681	1.681	1.681	1.681	1.681
Capex	$M	$131.10	$160.79	$155.81	$209.97	$276.27	$116.96
Return on Capital (ROC) (2,20,10)	$M/y	$18.14	$22.25	$21.56	$29.06	$38.23	$16.18
Opex	$M/y	$38.63	$14.51	$11.43	$13.68	$11.09	$9.65
Maintenance	$M/y	$4.04	$4.94	$4.78	$6.43	$11.86	$3.50
Utilities	$M/y	$125.11	$42.72	$34.40	$29.50	$42.85	$18.79
Total Cost	$M/y	$185.92	$84.42	$72.18	$78.66	$104.03	$48.12
	$/t CO_2	$110.61	$50.23	$42.94	$46.80	$61.89	$28.63

approaches. In a similar manner to that described for flue-gas extraction, this work has been translated into a 2018 price for a US Gulf site and is presented in Table 3.8.

Six methods are described. Diethanolamine (DEA), Selexol and triethanolamine (TEA) involve absorption, The Benfield process uses potassium carbonate solution as absorbent and focuses on acid-gases present. These case studies are supplemented by a membrane process and a cryogenic process. All these processes are assisted by the carbon dioxide being in relatively high concentration (typically 30%) and a high pressure (1 MPa). This increases the efficiency of the processes relative to flue gas and results, apart from the DEA process, which has an unusually high utilities requirement, reduces the cost of carbon dioxide extraction to the region of $50/t. Of the processes studied, the cryogenic process that produced liquid carbon dioxide was the lowest cost.

Carbon Dioxide Geo-Sequestration — Carbon Capture and Storage

Having extracted carbon dioxide from the hydrogen production process, it is delivered at a suitable pressure for transport by pipeline. This pressure is often very high to convert the gas into a dense phase or liquid at the pressure required for the method of disposal. The details of pipeline cost are discussed in Chapter 5, but for now a rule of thumb is that the capital cost of a pipeline will be in the vicinity of $1 million/km. Unfortunately for many sites, emitting large volumes of carbon dioxide, suitable sites for disposal are remote from the production site and long pipelines would be required. Suitable disposal sites fall into five general categories:

- Deep coal beds unsuitable for mining. The beds may be unsuitable because of salinity, sheer depth of operation required or steeply sloping coal seams. In this system, the carbon dioxide becomes adsorbed onto the coal. For this method, carbon dioxide has to be delivered at a pressure of typically 8–10 MPa.
- EOR can utilise carbon dioxide as a dense phase. This lowers the viscosity of heavy oils and increases their mobility of the oil in the oil-bearing strata. This is a method in common use in many parts of the world especially in North America where there is an active market for carbon dioxide for this purpose. As the dense phase carbon dioxide displaces the oil, it remains in the voids created. Any carbon dioxide that dissolves in the oil is separated and recycled to the well. For this purpose, the carbon dioxide has to be at a pressure of typically 15 MPa or higher.
- Saline aquifers are used in some current storage schemes.[23] The preferred locations lie below current oil or gas producing strata where it can be assumed that the deposited carbon dioxide will not subsequently leak back to the surface. For this purpose, the pressure of the carbon dioxide required is in the region of 15 MPa–18 MPa.
- Abandoned oil and gas wells are often targeted for disposal. There have been several trials and demonstrations of this method.[24] However, the issue of well integrity is important as there is an issue of unwanted migration of the carbon dioxide away from the disposal site. For this disposal option, the carbon dioxide is at a pressure of typically 15 MPa–18 MPa.

[23] For instance, the Gorgon project on Barrow Island in NW Australia.

[24] For example, the Otway geo-sequestration project in Victoria, Australia.

- Sea disposal in deep ocean basins or trenches below 1000 m in depth has also been proposed. This requires the access to these deep ocean fissures. The carbon dioxide pressure required is in excess of 30 MPa.[25]

Some idea of the cost for carbon dioxide compression have been provided by Saxena and Flintoff[26] for carbon dioxide available at near atmospheric pressure (e.g. from flue gas); Table 3.9.

Table 3.9: Estimates for compression power required for geo-sequestration of carbon dioxide from flue gas.

Carbon dioxide	t/y	500,000
	m^3/h	39,411
Suction pressure	kPa	115
Discharge pressure	MPa	18
Compression power	kW	6,610

Large-scale operations, such as those required for large-scale hydrogen production, would require the geo-sequestration of many millions of tonnes of carbon dioxide so that the power for the compression required for many geo-sequestration operations would represent a considerable cost burden.

Greenfield geo-sequestration

In the above discussion of possible geo-sequestration techniques, new greenfield developments have not been explicitly discussed. As many major emission points are remote from the preferred disposal sites there is the option of exploration and development activities aimed at finding suitable sites near to the point of emission. To-date very little has been done in this area because it would require a development program involving extensive new seismic studies followed by an extensive drilling program to delineate a prospective site.

[25] Caldeira K, Adams EE. (2009) Carbon sequestration via direct injection into the ocean. In: *Encyclopaedia of Ocean Sciences*, pp. 495–501.
[26] Saxena MN, Flintoff W. (December 2006) *Hydrocarbon Processing*, p. 57.

Another major problem is that for the large-scale disposal of carbon dioxide from a continually emitting point source, the disposal site would be required to continually absorbed the carbon dioxide at the same rate as the emissions are produced. The problem is in a sequestration well, as time progresses and the voids fill with gas, there will be increasing resistance to injection. For large disposals, extensive drilling in prospective strata will be required and potentially many hundreds of wells required over the life of the operation.

Finally, a new prospective site must have assurance that the site is viable in the long term (100 years or so) and that the acidic carbon dioxide (and water forming carbonic acid) will not result in dissolution of minerals (e.g. limestone) leading to leakage from the site.

Development Status

The Global Carbon Capture and Storage (CCS) Institute[27] has published a list of large-scale (>0.5 Mt/y CO_2) geo-sequestration projects in operation and under development. These are reproduced here in Tables 3.10 and 3.11. The tables show the commencement date or the planned date of commencement.

Of the operating facilities the majority are separating the carbon dioxide from relatively pure and concentrated streams (Sep.). Two are capturing carbon dioxide from flue gas — post-combustion capture (PCC). The major portion of the projects are using the captured carbon dioxide in EOR operations that might be operated by third parties remote from the extraction facility; Table 3.10.

For projects that are under construction, in development or in planning (Table 3.11), a major portion is focussed on EOR operations rather than dedicated disposal sites. The major portion of the projects are separating carbon dioxide from relatively pure and concentrated streams. Some of the projects link up many different users to build a critical mass for a CCS operation and there are consequently various (Var.) methods for obtaining the carbon dioxide.

---

[27] Global CCS Institute: www. globalccsinstitute.com.

Table 3.10: Operating large-scale carbon capture and storage (CCS) facilities.

Title	Status	Country	Date	Industry	Mt/y CO_2	Type	Storage
Gorgon CO_2 Injection	Operating	Australia	2019	Natural gas	3.4 to 4	Sep.	Dedicated
Jilin Oil Field CO_2 enhanced oil recovery (EOR)	Operating	China	2018	Natural gas	0.60	Sep.	EOR
Illinois Industrial CCS	Operating	USA	2017	Ethanol	1.00	Sep.	Dedicated
Petra Nova CCS	Operating	USA	2017	Power Gen	1.40	Post-combustion capture (PCC)	EOR
Abu Dhabi CCS (Phase 1)	Operating	UAE	2016	Steel	0.80	Sep.	EOR
Quest	Operating	Canada	2015	Hydrogen	1.00	Sep.	Dedicated
Uthmaniyah CO_2-EOR	Operating	Saudi Arabia	2015	Natural Gas	0.80	Sep.	EOR
Boundary Dam CCS	Operating	Canada	2014	Power Gen	1.00	PCC	EOR
Petrobras Santos Basin CCS	Operating	Brazil	2013	Natural gas	3.00	Sep.	EOR
Coffeyville Gasification	Operating	USA	2013	Fertiliser	1.00	Sep.	EOR
Air Products SMR	Operating	USA	2013	Hydrogen	1.00	Sep.	EOR
Lost Cabin Gas Plant	Operating	USA	2013	Natural gas	0.90	Sep.	EOR
Century Plant	Operating	USA	2010	Natural gas	8.40	Sep.	EOR
Snohvit CO_2 Storage	Operating	Norway	2008	Natural gas	0.70	Sep.	Dedicated
Great Plains Synfuels	Operating	USA	2000	SNG	3.00	Sep.	EOR
Sleipner	Operating	Norway	1996	Natural gas	1.00	Sep.	Dedicated
Shute Creek Gas Plant	Operating	USA	1986	Natural gas	1.00	Sep.	Dedicated
Enid Fertiliser	Operating	USA	1982	Fertiliser	0.70	Sep.	EOR
Terrel Natural Gas	Operating	USA	1972	Natural gas	0.40	Sep.	EOR

Table 3.11: Large-scale carbon capture and storage (CCS) projects in construction phase, under development or in planning.

Title	Status	Country	Date	Industry	Mt/y CO_2	Type	Storage
Alberta Carbon Trunkline	in const.	Canada	2020	Hydrogen	1.2–1.4	Sep.	Enhanced oil recovery (EOR)
Alberta Carbon Trunkline	in const.	Canada	2020	Fertiliser	0.3–0.6	Sep.	EOR
Sinopec Qilu Pet Chem CCS	in const.	China	2020	Chemicals	0.40	Sep.	EOR
Yanchang Integrated CCS	in const.	China	2021	Chemicals	0.41	Sep.	EOR
Wabsh CO_2 Sequestration	Adv. Devel.	USA	2022	Fertiliser	1.5–1.75	Sep.	Dedicated
Rotterdam CCUS	Adv. Devel.	Netherlands	2023	Various	2.0–5.0	Var.	Dedicated
Norway Full Chain CCS	Adv. Devel.	Norway	2024	Cement etc.	0.80	Var.	Dedicated
Lake Charles Methanol	Adv. Devel.	USA	2024	Chemicals	4.20	Sep.	EOR
Abu Dhabi CCS (Phase 2)	Adv. Devel.	UAE	2025	Natural gas	1.9–2.3	Sep.	EOR
Dry Fork CCS	Adv. Devel.	USA	2025	Power	3.00	Post-combustion capture (PCC) Sep.	Ded./EOR
Illinois Macon County	Adv. Devel.	USA	2025	Power and ethanol	2.0–5.0	PCC Sep.	Ded./EOR
Project Tundra	Adv. Devel.	USA	2026	Power Gen	3.1–3.6	PCC	Ded./EOR
Mid Continent CCS	Adv. Devel.	USA	2025	Various	1.90	Var.	Ded./EOR
Carbon Net	Adv. Devel.	Australia	2020s	Various	3.00	Var.	Dedicated
Oxy and White Energy	Planning	USA	2021	Ethanol	0.6–0.7	Sep.	EOR
Sinopec Eastern China CCS	Planning	China	2021	Fertiliser	0.50	Sep.	EOR
Hydrogen 2 (H@M)	Planning	Netherlands	2024	Power	2.00		Dedicated

From the perspective hydrogen production of some of the more interesting CCS operations are:

Great plains synfuels — Dakota gasification company (North Dakota, USA)

This is probably the most pertinent as it is an established commercial operation. In this operation that was first established as a result of high gas prices in the early 1980s, 6 Mt/y of lignite is gasified (Lurgi gasifiers) to produce synthesis gas that is subsequently converted into synthetic natural gas (153 MMscfd). Since 2000, both carbon dioxide (2.5–3.0Mt/y) and hydrogen sulphide have been extracted as a comingled stream and sent by a 320 km pipeline to the Weyburn field in Canada for EOR. The importance of this operation is that it clearly demonstrates all the major aspects of CCS required by a fossil fuel to hydrogen process.

Norwegian CCS — Sleipner gas field CCS and Snohvit CCS

The Sleipner West gas field in the Norwegian sector of the North Sea contains 9% carbon dioxide. In 1996, the world's first offshore CCS operation was established. The project uses an amine absorber technology on an offshore platform to reduce the carbon dioxide content of the gas below 2.5%. Approximately 0.9 Mt/y of gas is geo-sequestrated into a saline aquifer system contained in a 200 to 250 m thick porous sandstone formation (Utsira formation).[28] This aqueous system is thought to have an ultimate capacity of over 600 billion tonnes of carbon dioxide. Injection costs are reported at $17/t of carbon dioxide. The Sleipner gas field is now in its decline phase.

The Snohvit LNG is in the far north of Norway processing gas from the Barents Sea. The field has been developed entirely by sub-sea completions (no platform). The gas contains about 8% carbon dioxide which is

[28] Furre A-K, Eiken O, Alnes H, Vevatne JN, Kiaer AF. (2017). 20 Years of Monitoring CO_2-injection at Sleipner. *Energy Procedia* **114**: 3916–3926.

piped to the shore (160 km distant) where the carbon dioxide is extracted and transported back by another pipeline (153 km) to be injected into a large sandstone formation (Tubasan) 2.6 km below the sea bed.

These operations effectively demonstrate the availability and use of large reservoirs suitable for the disposal of large volumes of carbon dioxide and that the operation of the geo-sequestration can be performed remotely without the need for on-site facilities.

Gorgon gas project (Barrow Island, Western Australia)

The Gorgon Project uses gas from several fields to feed a major LNG facility on Barrow island offshore NW Australia. The facility has been exporting LNG since 2016. The island itself was once an oil producing area so the underground geology is well known. The raw gas contains carbon dioxide which has to be removed prior to LNG production. The carbon dioxide is piped to one of three drill centres on the island where the carbon dioxide is compressed and then injected into the Dupuy formation below the island. The aim is to geo-sequestrate 3.4–4 Mt carbon dioxide per year. Over the life of the project (40 years) some 100 Mt of carbon dioxide are expected to be geo-sequestrated.[29]

Case Study

CCS is central plank in the many schemes to develop a carbon emission free power network. There are implications for a hydrogen economy if an operation is to rely wholly or in part on power purchases from a central grid. This case study examines the implications of CCS in power generation and supply for large-scale power generation from a coal generator. The results can be used as a guide for large-scale CCS in the production of hydrogen from fossil fuels.

The case considered is for a 500 MW thermal power generator using coal as the fuel. This is a typical size although many new operations are larger (2,000 MW), essentially comprise multiple modules of smaller

[29] Chevron Australia Pty. Ltd. "Gorgon Carbon Dioxide Injection Project" australia.chevron.com, 2009.

generators. There may be some economies of scale that have not been considered in this analysis. The following assumptions are made about the generator:

(i) The generator operates with load factor of 80%, so the yearly output is 400 MW with an annual output of 3.5 millionMWh (12.6 PJ/y).
(ii) The generator operates with a thermal efficiency of 39%.
(iii) The coal input is high grade thermal coal with:
- A specific energy of 29.71GJ/t,
- A carbon content as received of 74.6%
- A sulphur content of 0.78%.

(iv) The flue gas contains 15% carbon dioxide and the extraction system reduces this to 2%, i.e. an extraction efficiency of 13/15 = 87%.
(v) The carbon dioxide produced is 8,155 t/d and 7,095 t/d of carbon dioxide is extracted, leaving 1,060 t/d in the flue and discharged to atmosphere.

The downstream units for carbon capture are illustrated in Figure 3.12.

Waste-heat recovery

The flue gas is assumed to be at 150°C and at atmospheric pressure. The duty of the waste-heat recovery system is to reduce the temperature to ambient and remove excess water in the gas. It is envisaged that this could be accompanied by raising low-pressure steam but the bulk of the cooling will be by means of water sprayed into the flue gas. This would wash out

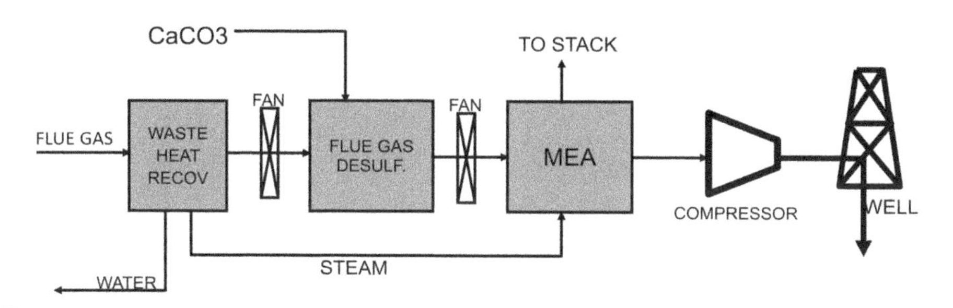

Fig. 3.12: Downstream units considered in carbon capture and storage (CCS) case study.

any particulates in the gas and would require a large quantity of cooling water estimated at about 36,000 ML/y. This will be recirculated so the cost is for make-up cooling water (5% of the requirement). Make-up water is costed at $1000/ML.

Fan (compressor)

It is necessary to increase the pressure of the flue gas to facilitate its movement through the following stages. For this study, it is assumed that this can be accomplished by means of a fan-blower type system lifting the gas pressure to 150 kPa (1.5 bar). As envisaged in the diagram, this is accomplished in two parts, one before and one after, sulphur extraction.

Flue-gas desulphurisation

This is considered necessary to reduce solvent loss in the downstream unit. This is accomplished by passing the gas through crushed limestone or in the form of a limestone slurry. Assuming this is 100% efficient, 26kt/y of limestone would be required. Limestone is costed at $30/t.

MEA carbon dioxide extraction

A standard MEA system is considered similar to that described above. It is assumed that heat extracted from the waste-heat recovery is sufficient for the solvent regeneration system. As noted above, the extraction is assumed to be 87% efficient.

Compression, pipeline and CCS wells

The duty of the compressor is to compress the carbon dioxide to produce liquid at a sufficient pressure (18 MPa) for transmission by pipeline (200 km) to the disposal wells.

The outcome of this case study is given in Table 3.12:

The capital cost is dominated by the cost of increasing the pressure of the flue gas, the MEA solvent extraction system and the pipeline. Ancillaries (unspecified pumps, piping etc.) are at a cost of 15% of the

Table 3.12: Statistics for carbon capture and storage (CCS) case study.

Section	CAPEX	OPEX	Power	Cooling water	Power
	$M	$M/y	$M/y	$M/y	MW
Waste-heat recovery	$10.00	$1.00		$1.81	
Fan-blowers 1 & 2	$119.55	$11.96	$7.34	$0.49	18.00
Flue-Gas Desulphurisation	$20.00	$2.99			
Methyl ethanolamine (MEA)	$186.68	$29.38	$14.88	$4.29	33.98
Compressor	$76.30	$7.63	$14.94		34.10
Pipeline	$200.00	$10.00			
Wells	$50.00	$5.00			
Sub total	$662.54	$67.96	$37.16	$6.59	86.08
Ancillaries (15%)	$99.38	$9.94			12.91
Total	$761.92	$132.91	$66.98	$10.87	99.00
	$M/y				
Return on Capital (ROC) (2y const.; 20y life; 10% DCF)	$103.37			%OUTPUT	24.7%
Total operating costs	$210.77				
TOTAL	$314.13				
Unit CO_2 cost	$91.46				

unit costs to give a total capital estimate for this CCS system of $760M. Operating costs of the individual unit operations reflect the capital cost (typically 10% of Capex plus solvent, limestone etc.). power cost is assumed to be at $50/MWh.

The result of the estimate is that the cost for removal and geo-sequestration of 87% of the carbon dioxide in the flue gas is about $90/t. Importantly, the operation will consume 99 MW or nearly 25% of the generators output so that it is a reasonable assumption that this will be reflected in an increase in power cost of this level.

Discussion

With the precision of this high-level analysis, the result of this case study is very similar to the estimates provided by the earlier analysis presented above, namely the cost of CCS of carbon dioxide from flue gas is about $100/t of carbon dioxide. This is likely to be an optimistic estimate since the costs associated with defining and operating the geo-sequestration sites has not been considered.

The cost of carbon dioxide capture from more concentrated and cleaner sources is significantly lower at between $30 and $50/t of carbon dioxide. Again this does not include the cost of defining and operating the geo-sequestration site. However, in this instance, the costs are comparable to the known purchase cost for carbon dioxide by companies engaged in EOR schemes.

For the cases following, it is assumed that carbon dioxide CCS cost are $50/t from high concentration and clean streams. Carbon capture from flue gas schemes is not considered unless otherwise noted. Rather the emission cost is assumed to be satisfied by the purchase of carbon credits at $50/t unless otherwise noted.

CHAPTER 4

THE PRODUCTION OF HYDROGEN FROM RENEWABLE SOURCES

One of the generally unstated aims of promoters of the hydrogen economy is that the hydrogen should be produced entirely by so-called renewable sources and in particular wind-generated electricity, hydropower, solar voltaic cells, renewable biomass etc. The purpose of this chapter is to review these methods and identify the hurdles and costs of implementation by these routes.

Although the generation of hydrogen on a large scale and export to a central fuel filling facility is an important objective there is also the objective of producing hydrogen on a small scale to fuel domestic requirements such as fuel cells for power generation. This is also considered in this chapter.

For the most part we are considering the use of renewable electricity as the source of hydrogen production by electrolysis so this is considered first.

The Production of Hydrogen by Electrolysis

The passage of electricity through water results in the dissociation of the water molecules into hydrogen and oxygen.

$$\mathbf{2H_2O = 2H_2 + O_2} \qquad \textbf{(4.1)}$$

So that for every two moles of hydrogen produced a mole of oxygen is also produced and two moles of water are required. Pure oxygen is dangerous so that for the production of hydrogen, especially on a mass scale, the oxygen has to be disposed of by discharge to atmosphere in a safe manner or utilised in some way. Furthermore, the requirement for two moles of water (of high purity) can be an issue in remote areas which may be the optimum location for solar farms for electricity production.

Electrolysis

The fundamentals of water electrolysis were developed in the 19th and early 20th centuries. The basis of the process is illustrated in Figure 4.1. Two metal electrodes are connected by a direct current (DC) supply to form an anode and a cathode. The current is passed through the solution by hydrogen ions passing from the anode to the cathode and hydroxide ions passing from the cathode to the anode. At the cathode the hydrogen ions are neutralised by electrons delivered by the electric circuit to form hydrogen gas, which forms as bubbles around the cathode and is discharged out of the cell.

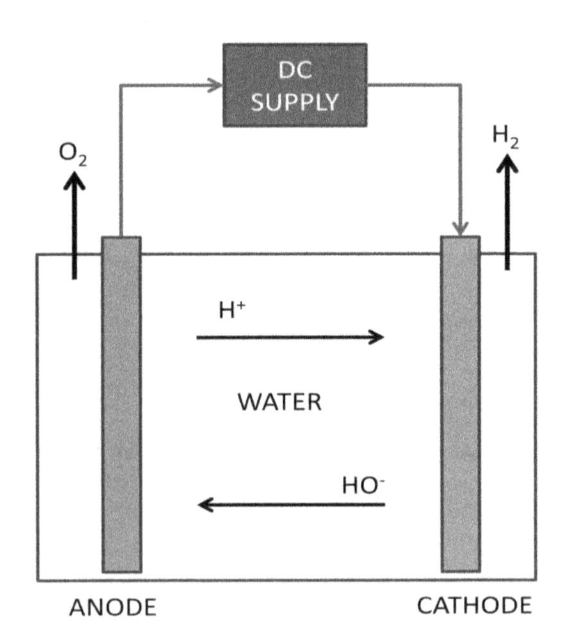

Fig. 4.1: Water electrolysis.

$$2H^+ + 2e^- = H_2 \qquad (4.2)$$

Simultaneously, hydroxyl ions are attracted to the anode liberating oxygen and electrons which flow round the circuit to the cathode.

$$4HO^- = O_2 + 2H_2O + 4e^- \qquad (4.3)$$

This is the basic mechanism of water electrolysis but, as is commonly the case with simple explanations, each part of the process is more complex.

Overview of Cell Operation

Water

Water is a poor electrical conductor with a low rate of ionisation into hydrogen and hydroxyl ions. This seriously restricts the rate of ion transfer in the water which determines the rate of hydrogen production and which is proportional to the current (A) passing through the cell. Fortunately, the ion transfer can be boosted by dissolving ionic materials in the water — acids, bases or salts. The use of salts is problematic in that, depending on the position of the anion and cation (the negative and positive ions) in the electrochemical series, materials other than hydrogen and oxygen can be liberated. For instance, the presence of copper can result in deposition of copper on the cathode rather than hydrogen liberation. This is of course useful for copper refining when a copper salt is electrolysed to deposit pure copper on the cathode. It is also an issue with chloride salts since when the chloride anions (the negative ions) are attracted to the anode, highly toxic chlorine can be liberated or explosive chlorate salts formed. Since salt (sodium chloride) is commonly found in water supplies, this means that the water to be used has to be treated to remove salts which would otherwise compromise the liberation of oxygen and hydrogen.

The preferred ionic media to add are strong (in the sense of high levels of ionisation) acids but preferably bases, as the latter are not volatile and not lost to the system. In particular, the hydroxides of the alkali metals are commonly used. The most commonly used electrolyte is potassium hydroxide at a concentration of about 30% and this type of cell is generally referred to as an alkaline cell. Adding these materials greatly increases the ion transfer rate and boosts hydrogen and oxygen output. However, these materials change the basic equations detailed above.

When acids are used, the transfer of charge in the solution is dominated by the movement of hydrogen ions from the anode to the cathode; "See the reaction mentioned in Eq. 1. The liberation of oxygen at the anode involves the break-up of water molecules:

$$2H_2O = 4H^+ + O_2 + 4e^- \quad (4.4)$$

Similarly, when a base is used to boost the current density, the main charge transfer agent is hydroxyl ions and as a consequence, the main hydrogen forming reaction at the cathode is

$$2H_2O + 2e^- = H_2 + 2HO^- \quad (4.5)$$

There can be variations on this, e.g. by introducing a membrane to split the electrolytic cell the anode and cathode can be operated with different electrolyte compositions (Figure 4.2).[1]

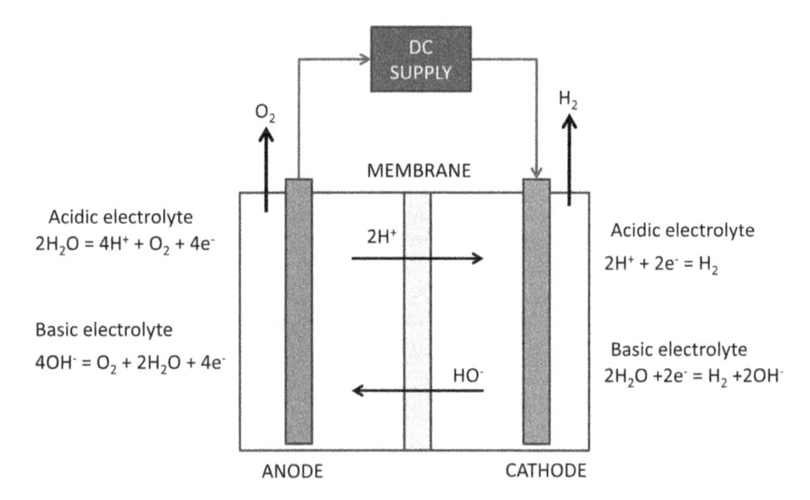

Fig. 4.2: Electrolysis cell with membrane to facilitate either acid or base electrolytes.

[1] Gomez RMA. (5 February 2008) *US Patent* 7,326,329 and (11 January 2019) *US Patent* 10,316,416 discusses membrane split cells and proposes variants for mass hydrogen production.

The cathode

The generation of hydrogen with an acidic electrolyte simply requires the neutralisation of hydrogen ions followed by coupling to produce hydrogen molecules. As hydrogen ions gather around the cathode there is some resistance to the process which creates an overvoltage, which is when the voltage to discharge hydrogen is higher than ideal. On platinum metal this is a facile process to overcome and platinum is often the optimum choice for inclusion in the cathode. The use of noble metals in the cathode clearly increases the capital cost of electrolytic cell.

The anode

The reactions at the anode are more complex than those at the cathode as they involve the multiple steps to produce oxygen gas. This complexity results in an electrode polarisation and larger overvoltages. The choice of the ideal electrode material is less clear. Often transition metals such as nickel are used. These metals oxidise to form a passive layer that prevents dissolution of the anode. It is possible that the partially oxidised anode participates in the process of oxygen evolution by acting as an oxygen transfer agent.

The DC supply

The theoretical minimum voltage (EMF) that is required to electrolyse water into its elements under ambient conditions is 1.23 volts. However, in practice, the voltage required is typically about 1.9 volts. This overvoltage difference is a consequence of the polarisation phenomena particularly at the anode. Overvoltages are exacerbated by high current densities that place a limit to the output of any particular cell. Typical cells operate at current densities of about 0.5 A/cm^2 for alkaline cells and somewhat higher for proton exchange membrane (PEM) cells (the various types of cell are discussed below). The voltage at which hydrogen and oxygen are formed is known as the discharge voltage and if the voltage applied is less than the discharge voltage, no hydrogen is produced. As the voltage required is directly related to the energy required (volts × amps = watts) and hence the operating cost of the cell, it is important to minimise the discharge voltage.

Another point to note is that the overvoltage represents the energy inefficiency of the system, so if we take the energy efficiency as the ratio of the theoretically ideal voltage to the actual voltage we see the efficiency of the cell is about 65%. Fuel cells, which are discussed later, also suffer from the overvoltage problem, which limits their energy efficiencies in a similar manner to about 60%. In passing, it is interesting to note that the overall efficiency (from hydrogen generation to vehicle use) will be a product of these efficiencies, namely about 39% which is not too dissimilar to the efficiencies achieved by modern internal combustion engines.

Impact of temperature on required EMF

The EMF (voltage) of electrolysis is determined by considering the fundamental thermodynamics of the process. The required EMF[2] is determined by the free energy[3] (G) change for the reaction:

$$\mathbf{H_2O_{(l)} = H_{2(g)} + 1/2O_{2(g)}} \tag{4.6}$$

where the reactants and gases are in their standard states at 25°C. Note that water is in the liquid state and the free energy ($\Delta G°$) of change on vaporisation is added to the free energy of change of the reaction, which is

$$-\Delta G° = zFE° = \mathrm{R}T\ln\mathrm{K} \tag{4.7}$$

and

$$E° = -\Delta G°/zF \tag{4.8}$$

where $E°$ is the standard EMF, R is the universal gas constant (8.3144 J/K/mol), T the absolute temperature (298 K), lnK the natural logarithm of the equilibrium constant(K), z the charge on the ions (unity) and F the Faraday (96,487 coulombs/electron mole). The influence of temperature on the EMF required (E) is determined from the Nernst equation, namely:

[2] The electrochemistry of cells is discussed in physical chemistry texts.

[3] Free energy (G) is a thermodynamic function and is a combination of the enthalpy (H) and the entropy (S) by $G = H - TS$ where T is the absolute temperature.

$$E = E^\circ - (RT/zF)\ \ln K \quad (4.9)$$

As the temperature rises, the required EMF falls so that at 1000 K this is estimated at 0.99 volts versus E° at 1.23 volts at 25°C.

Equations of this type (Nernst equation) are common in understanding the basic thermodynamics for operation of electrolysis cells, fuel cells and membrane compressors.

Impact of pressure on required EMF

Since the pressure of the system influences the equilibrium constant (K in Eq. 4.9) changing pressure will influence the EMF. Increasing pressure increases the required EMF. The standard EMF (E°) will rise from 1.23 volts at atmospheric pressure to about 1.26 volts at 1 MPa and 1.29 volts at 10 MPa. Note these are ideal voltages and in practice will be higher.

Membrane Compression of Hydrogen

At this point it is pertinent to discuss the relatively new technology of membrane compression of hydrogen. This is finding increasing use in the small-scale compression of hydrogen produced from relatively small electrolysers and avoids the high cost of axial compressors which are used in the large-scale compression of hydrogen gas. Figure 4.3 illustrates the general layout of the system.[4]

In a membrane compressor, a membrane is sandwiched between an anode and a cathode. The membrane comprises a polymer electrolyte membrane capable of transporting hydrogen ions (protons). These PEM are constructed from perfluoro polymers[5] with acidic groups as

[4] Wong TYH, Girard F, Vanderhoek TPK. (28 October 2004) *US Patent Application* 2004/02116679, to NRC Canada. This paper gives a good explanation of the theoretical and practical workings of the system.

[5] Perfluorocompounds such as those used to construct proton exchange membranes are considered by the Stockholm Convention on persistent organic pollutants (POPs) as an emerging pollution issue because they breakdown in the environment to perfluorinated alkyl acids and alcohols which are deemed to have major environmental impact. The aim

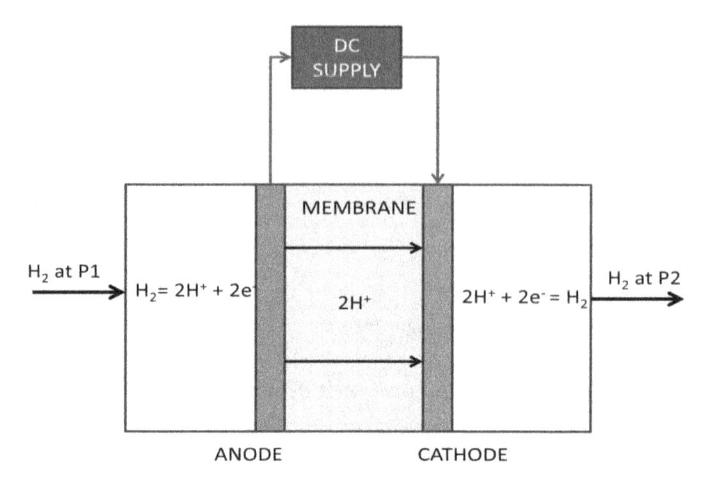

Fig. 4.3: Membrane compression.

appendages such as Nafion™. The anode and cathode are typically made from platinum supported on graphite. Hydrogen enters the anode compartment at a low pressure, e.g. from water electrolysis at atmospheric temperature. A DC current is applied to the cell that causes the hydrogen to form hydrogen ions which diffuse through the membrane to the cathode where the hydrogen ions are neutralised reforming hydrogen gas at a higher pressure.

The Nernst equations are applicable, which for this application takes the approximate form:

$$\Delta E = (RT/2F)\ln\left(P_{H2out}/P_{H2in}\right) \tag{4.10}$$

where ΔE is the applied potential and P is the pressure of the gas.

In practice, the EMF required is higher than the theoretical value due to:

- The Ln (Pout/Pin) term deviate at high pressures due to hydrogen not behaving ideally at high pressure.
- Over-potentials arising from resistance across the anode, cathode, the proton conductive separator and the electric circuit resistance.

---

of the Convention is to limit their use with the eventual aim of phasing out these materials.

- Backflow of hydrogen from the high-pressure side to the low-pressure side.

There are reports that the efficiency of the system can be as high as 80% but in practice the efficiency may be lower, in the order of 50% or less.

To attain high pressures suitable for vehicle charging and to help minimise the over-potentials and back-flow, multiple units are added in series with the first cell feeding gas into a second cell etc. In this manner it is claimed the units can increase the pressure from atmospheric to over 8 MPa, which is suitable for vehicle charging. As one may deduce, building a unit with one side at atmospheric pressure and the other at 8 MPa will create many stresses in the system resulting in leaks and O-ring failure, which is an impediment to constructing a durable unit.[6]

High-Temperature Electrolysis

The PEM introduces the idea of membranes selective for oxygen rather than hydrogen. There exists a series of such membranes based on ceramics made from combinations of yttria (Y_2O_3) and zirconia (ZrO_2). Placing yttrium atoms (valence number 3) into an oxide matrix of zirconium (valence number 4) naturally introduces defects in the structure. At high temperatures (i.e. temperatures expected to be vibrationally, in the chemical sense, excited), oxygen can be captured by the surface defects and absorbed into spaces within the structure as oxygen anions. A pressure potential will allow the oxygen anions to shuttle across the ceramic and will allow reformation of oxygen on the other side of the ceramic by oxygen anion recombination. Such ceramics are being developed for the conversion of natural gas solid oxide fuel cells and for high-temperature electrolysis of water. Figure 4.4 illustrates the technology.

The cell is operated at about 800°C. Superheated steam enters the cathode side of the cell where an electrode potential induces absorbed water molecules to dissociate into hydrogen gas and oxide ions.

---

[6] Ukai K, Ejima T, Kawabata N. (23 April 2020) *US Patent Application* 2020/0124039, to Panasonic Intellectual Property Management Co. Ltd.

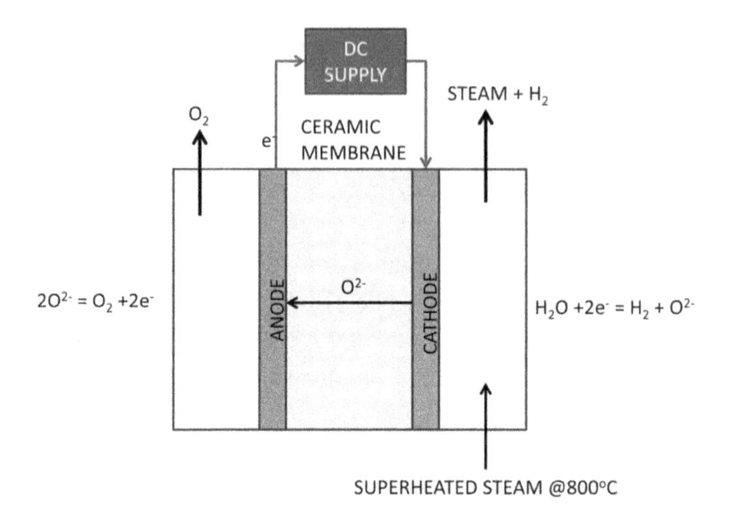

Fig. 4.4: High-temperature electrolysis.

$$H_2O + 2e^- = H_2 + O^{2-} \tag{4.11}$$

The oxide ions form on the ceramic membrane's surface, pass through it, and recombine at the anode to form oxygen and release electrons.

$$2O^{2-} = O_2 + 2e^- \tag{4.12}$$

The electrons pass round the circuit and the oxygen is discharged. The hydrogen is discharged with excess steam from the cathode side of the cell.

From the discussion above it is to be expected that electrolysis at high temperature requires a considerably lower voltage than at ambient temperature. Furthermore, the excess high-pressure superheated steam and hydrogen exiting the cell can be used to generate power with a train of turbo-generators and a final condensing generator to obtain water and separating the hydrogen. If this is part of a nuclear power generator cycle, the hydrogen with zero carbon emissions can be obtained as proposed in Figure 4.5.[7]

[7] O'Brien J. (2014) High temperature electrolysis for efficient hydrogen production from nuclear energy. In: *Electrolytic Hydrogen Production Workshop*, NREL, Golden Colorado (27–28 February 2014).

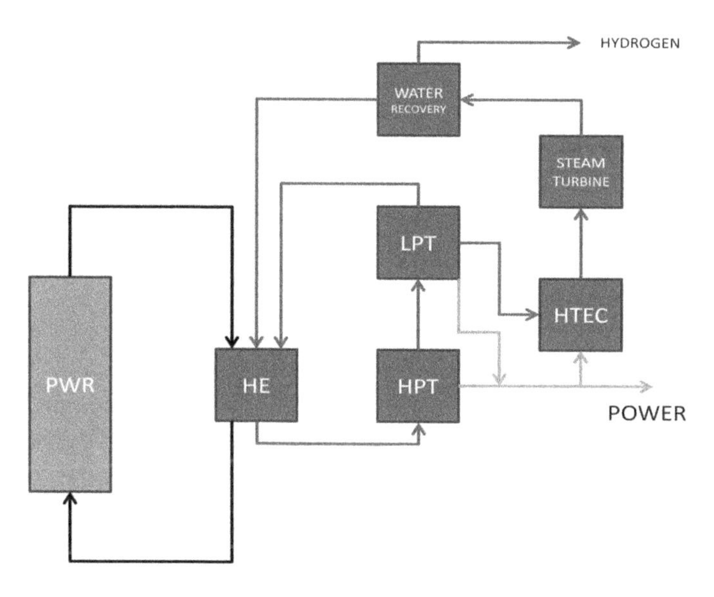

Fig. 4.5: Hydrogen production at high temperature using nuclear power.

In this scheme a pressurised water reactor (PWR) provides heat that is transferred to a steam system by heat exchange (HE). High-pressure superheated steam is passed to high-pressure turbine (HPT), which generates electrical power. Exhaust steam at lower pressure is passed to a low-pressure turbine (LPT). Exhaust steam from this unit passes to the high-temperature electrolysis cell (HTEC) which generates oxygen (not shown) and hydrogen which remains in the excess steam. This is passed to a steam turbine system that condenses the steam and allows hydrogen recovery from the water recycle unit.

A particular problem with the high-temperature electrolyser is the high-temperature steam attack on the cathode, which is typically nickel supported on a zirconia ceramic. Note that essentially solid oxide electrolysers are the same as solid oxide fuel cells but the erosion is particularly of concern when the unit is operated in the electrolysis mode. The erosion can be ameliorated somewhat by having hydrogen present in the input steam, which maintains the cathode in a reduced state.[8]

[8] Gallo PL. (27 November 2012) *US Patent* 8,317,986, to Commissariat a l'Energie Atomique.

Comparison of Performance of Different Cells

The above narrative describes the three types of electrolysis cells in general use — alkaline, PEM and solid oxide electrode (SOE) cells. The various types are compared in Table 4.1.

These cell systems comprise a cluster of individual cells in a stack. The stacks have generally limited lifetime (typically about 10 years) and are replaced as a whole.

Alkaline cells have been in the longest use and have seen the longest development phase. Relative to other types they can be large and have good durability (20-year life). They can be operated up to a pressure of about 3 MPa (30 bar), not enough for vehicle filling.

Some PEM systems claim to deliver hydrogen over 15 MPa (150 bar) pressure which is still not enough for optimal vehicle charging (40 or 80 MPa is generally considered to be required). Energy consumption is about 10% higher than the equivalent alkaline cell. A major advantage of the

Table 4.1: Comparison of alkaline, proton exchange membranes and solid oxide electrolysis cells for the production of hydrogen (after Hinkley *et al.*[9]).

		Alkaline	PEM	SOE
Electrolyte		KOH solution	Proton exchange membrane	Ceramic
Electrode		Ni/Fe	Pt, Ir	Ni/ceramic
Operating temperature	°C	50–80	Ambient to 90	700–1000
Pressure (max)	Bar	30	165	1
Unit size (commercial)	MW	3.2	1.5	1
Hydrogen production	Nm^3/h	760	285	1
Current densities (max)	A/cm^2	0.6	2	Unknown
Energy consumption	kWh/kg	50–78	50–83	35.1
Lower partial load		20%–40%	0%–20%	Unknown
Cell lifetime	y	20+	<20	Unknown
Stack life (max)	h	90,000	80,000	Unknown
Capital cost	$/kW	850–1,500	1,500–3,800	Unknown

[9]Hinkley J, Hayward J, McNaughton R, *et al.* (21 March 2016) *Cost Assessment of Hydrogen Production from PV Electrolysis*, CSIRO, ARENA Project A-3018.

PEM system in the renewable sector is that the partial load range is very low or zero. As they are relatively new, the long-term durability is unknown. A downside is the present requirement to use noble metals in the electrode formulation.

The SOE systems are still in development. They are targeted for the high-temperature electrolysis systems as discussed above. A major drawback is their requirement for a low operating pressure difference across the cell. The cost of these systems is not yet known.

Other Approaches

The main problem with the conventional approach to the electrolysis of water to produce hydrogen is the high over-potential at the anode. This is a consequence of the multiple steps in the process involving oxide ion adsorption to the anode surface, oxide ion neutralisation and combination with another oxygen atom followed by desorption from the surface. This multi-step process inevitably increases kinetic barriers, which have to be overcome. Several alternative processes aim to solve this problem by eliminating oxygen liberation at the anode and replacing it with an alternative reaction chemistry.

One alternative approach is to divide the chemistry of water splitting into a thermal and an electrical operation. Such a thermo-electric cycle is based on sulphuric acid as the electrolyte.[10] In the cell reaction, the cathode reaction remains the same (Eq. 4.2) whilst for the anode the reaction becomes:

$$\mathbf{SO_2 + 2H_2O = H_2SO_4 + 2H^+ + 2e^-} \qquad \mathbf{(4.13)}$$

and for the cathode

$$\mathbf{2H^+ + 2e^- = H_2} \qquad \mathbf{(4.14)}$$

The theoretical EMF required for this process is 0.17 volts compared to 1.23 volts for the ideal electrolysis of water liberating oxygen. Sulphuric acid is withdrawn from the system and thermally decomposed as:

[10] Nakagiri T. (25 August 2009) *US Patent* 7,578,922, to Japan Nuclear Cycle Development Institute.

$$2H_2SO_4 = 2SO_2 + 2H_2O + O_2 \tag{4.15}$$

This approach would be attractive in locations with availability of cheap excess heat at a suitable temperature.

Another approach is the electrolysis of aqueous methanol. In this system, the cathode reaction remains the same whilst the anode reaction becomes the oxidation of methanol to carbon dioxide:

$$CH_3OH + H_2O = CO_2 + 6H^+ + 6e^- \tag{4.16}$$

The theoretical EMF for this reaction is about 0.4 volt. An advantage for this approach is that electrolysis of aqueous methanol can potentially be used to supply on-board hydrogen for vehicles.[11]

Impact of Solar Radiation Variation on Electrolysis

An important target for the hydrogen economy is the use of renewable electricity to produce green hydrogen. The renewable electricity source is usually considered to be wind or solar or a combination of the two. Solar is considered as a potentially major source in regions with high levels of solar irradiance, where very large solar arrays to produce power are being developed. A significant problem with this approach is that the solar irradiation is quite variable in areas which would naively be considered as having a stable solar flux. The problem is illustrated in Figure 4.6, which details the output of a PV (photo-voltaic) plant in central Australia over a 24-hour period.

As can been seen from the figure, there is considerable variation in output over very short time periods. If this output power is fed into a hydrogen electrolyser system, then the electrolytic cell would have to respond to this variation. This requirement promotes the use of PEM electrolysis cells over alkaline and other types because of their faster response time. Of course, the PV output could be smoothed using some sort of battery system but only at higher capital cost.

---

[11] Narayanan SR, Chun W, Jeffries-Nakamura B, Valdez TL. (6 June 2006) *US Patent* 7,056,428, to California Institute of Technology; Ibid., *US Patents* 6,432,284; 6,368,492; 6,299,744.

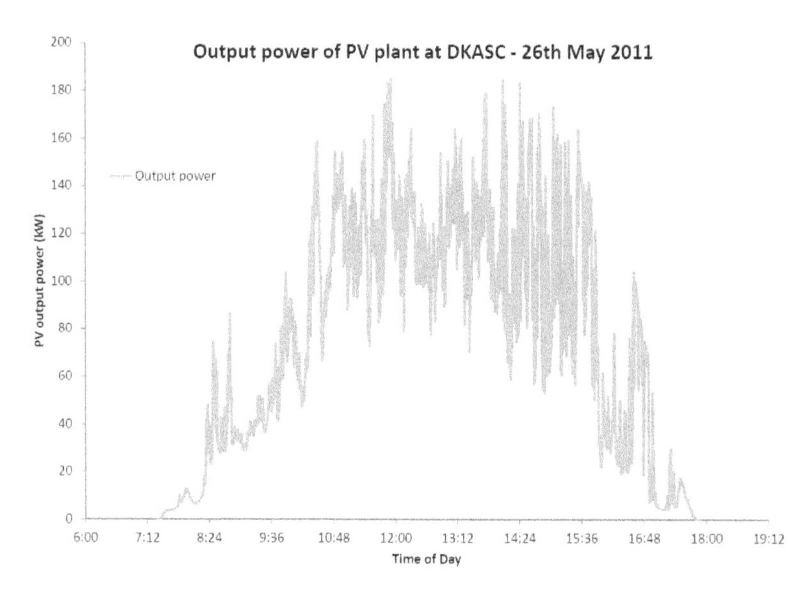

Fig. 4.6: Illustrative output from a photo-voltaic (PV) array in central Australia for 26th May, 2011.[12]

The efficacy of PEM systems in coping with this challenge is depicted in Figure 4.7, which illustrates the output achieved over a 5-day period from a solar array coupled to a PEM electrolyser. The figure illustrates that the output tracks the solar radiation curve well over the period which includes some periods where there is a rapid fall in radiation.

Economics of Hydrogen Production by Electrolysis

The basic equations for the reaction:

$$H_2O + 2e^- = H_2 + O^{2-} \quad (4.17)$$

are:

- 2 Faradays (F, 2 × 96,486 Coulombs(C)) generates 1 mole (2 g) of hydrogen

[12] Sayeef S, Heslop S, Cornforth D, *et al.* (2012) *Solar Intermittency: Australia's Clean Energy Challenge*. Desert Knowledge Australia Solar Centre.

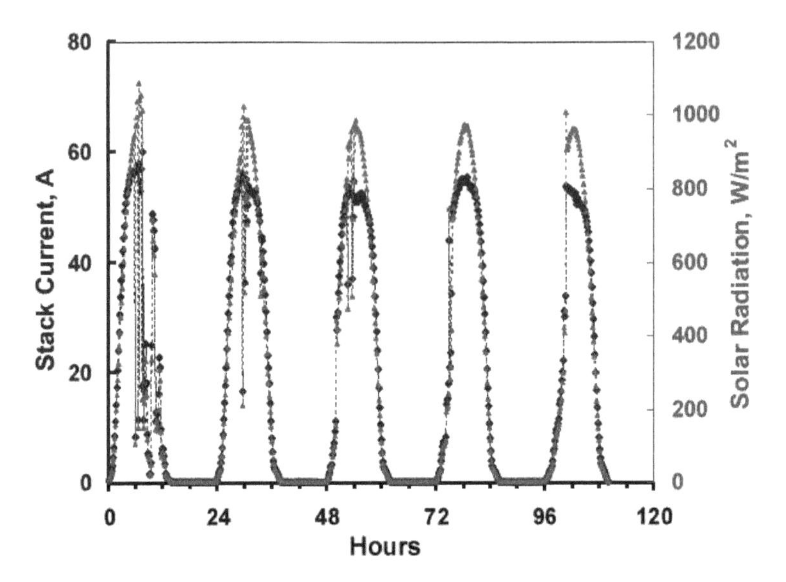

Fig. 4.7: Stack current and solar radiation data for a proton exchange membrane electrolyser over a 5-day period.[13]

- 1 Coulomb/sec = 1 Amp (C/s = A)
- At an EMF of 1.9 volts (V)

  1 mol H2/sec requires 1.9 × 2 × 96,486 VA (W)

  1 mol H2/sec requires 366,646 W or 366.65 kW
- Hydrogen production in 1 hour is 7.2 kg (2 × 3600/1000) using 366.65 kWh
- Hydrogen production rates are

  19.64 g/kWh

  9.82 mol/kWh

  0.232 Nm^3/kWh

  4.31 kWh/Nm^3

  50.92 kWh/kg

Production costs by electrolysis

Given the known cell efficiencies, the key variables for determining the production cost for such a facility are the capital cost and the cost of power.

[13]Clarke RE, Giddey S, Ciacchi FT, *et al.* (2009). Direct coupling of an electrolyser to a solar PV system for generating hydrogen. *Int J Hydrogen Energy* **34**: 2531.

Capital costs

A central production facility envisages the production of 373 kt/y of hydrogen. At this time, it seems likely that such a facility would comprise a grouping of smaller modules of cells rather than the expansion of a single large module, that is, there is limited economy of scale to the accessed.

Ulleberg and Hanke[14] have described the economics of small-scale operations in Norway using alkaline cells. Their data shows a fall in costs with scale, which is illustrated in Figure 4.8.

The figure illustrating the production costs shows a good economy of scale to about 1000 kg/d (1 t/d) after which there is little improvement in economy. Using the values given in the cost of production curve gives an estimate at the scale required generates negative values for the capital cost. However, the cost of production curve appears to reach an asymptote of about \$1,500/kg/d. Using this value, the capital cost for a large central facility operation is estimated to be \$1,650 M.

The conventional wisdom is that alkaline cells are cheaper than PEM cells of a similar capacity.[15] The Norwegian data illustrated is for alkaline

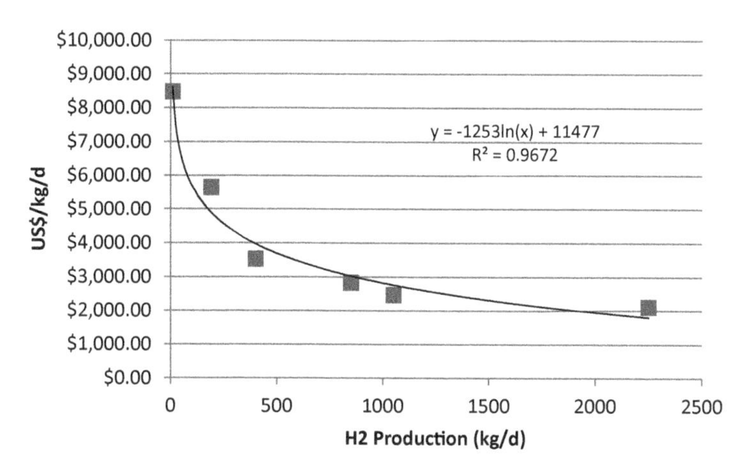

Fig. 4.8: Norwegian costs for small-scale hydrogen production.

[14] Ulleberg O, Hancke R. (2020) Techno-economic calculations of small scale hydrogen supply systems for zero emissions transport in Norway. *Int J Hydrog Energy* **45:** 1201–1211. Data taken from Figure 2; the original Norwegian Kroner data have been transcribed into US dollars at the rate of NOK/US\$ = 0.85.

[15] S&P Global Platts, "Methodology and specifications guide — Hydrogen" April 2020 uses a cost ration of alkaline/PEM of 78%.

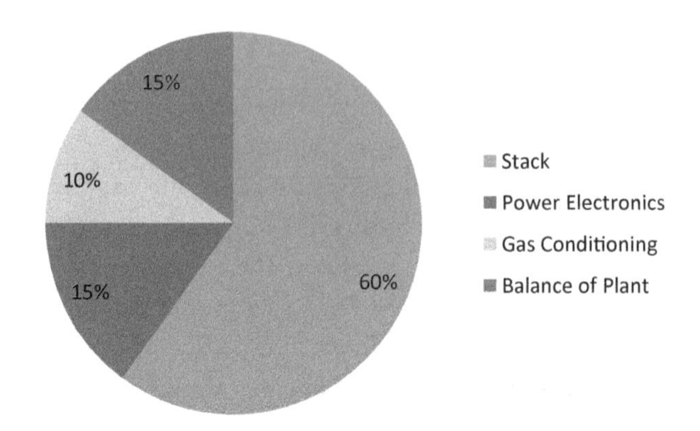

Fig. 4.9: Breakdown of electrolyser costs.

cells. However, despite reported higher costs for PEM cells, cost quotations in the Norwegian study for PEM cells where little different from the alkaline cell quotations.

Hinkley *et al.*[16] collated the published costs in $/kW for PEM electrolysis cells from several sources. Averaging and escalating the capital cost to 2019 values gives a capital estimate for a 3 MW system at $1,865/kW. If an economy of scale factor of 0.85 is achieved on the scale up to the size required for a central facility, the estimated capital cost is $M1,775, which is close to the estimate obtained from the Norwegian data.

One of the main fiscal issues relating to electrolyser cost is the life of the electrolyser stack. A typical breakdown of the capital cost is illustrated in Figure 4.9, which would seem to indicate that the replacement of the stack represents a large portion of the capital cost. Furthermore, the stack lifetime estimate of this seems to vary with source and it typically lies in the range 50,000 to 90,000 hr.

However, the regular replacement of 60% of the capital cost of the cell is probably an overestimate.[17] These capital estimates should be regarded

[16] Cost assessment of hydrogen production from PV and electrolysis. Jim Hinkley, Jenny Hayward, Robbie McNaughton, Rob Gillespie (CSIRO) Ayako Matsumoto (Mitsui Global Strategic Studies Institute) Muriel Watt, Keith Lovegrove (IT Power) 21 March 2016 Report to ARENA as part of Solar Fuels Roadmap, Project A-3018.

[17] In S&P Global Platts "Methodology and specifications guide — Hydrogen" April 2020 allows for a 15% of capital expenditure every 7 year (about 56,000 hr), which presumably is to allow for replacement of the electrolyser cell stacks.

as the cost of the Inside Battery Limits (ISBL) plant and the associated utilities so that for the large-scale central facility the cost of land and some other off-sites should be added.

Power costs

Clearly with such cell efficiencies — typically 65% for alkaline cells and 58% for PEM cells — the economics of hydrogen production would be very sensitive to the power price. The power purchased could be from a local grid system and would be a mix of fossil fuel generators, nuclear plant and renewable sources depending on the location. For these operations, the cost of power can be easily estimated from published sources such as regulatory bodies and in many cases the embedded carbon emissions from the mix of power going to the grid can also be derived.

Of significant interest is the powering of electrolysers using solar cells. At this time, there is limited data on the cost of power from solar arrays. Recent reports place this in the region of $68/MWh.[18] The non-power operating costs for both alkaline and PEM cells are estimated at 5%. Another approach to using renewable electricity, which could deliver lower power costs, is to use wind power which in many instances can deliver electricity well below the cost of solar generation. This could be used in combination with hydropower to deliver constant power and enable the use of more efficient electrolysis cells.

Estimate of hydrogen production costs by PEM electrolysis cell

The estimated cost for the production of hydrogen in a central facility based on a series of PEM electrolysis cells is given in Table 4.2. The efficiency of the system is 58%. At this efficiency, the 373 kt/y of hydrogen required for a central facility would require 2,628 MW, which is of course, a very large power demand for power to drive about 40,000 vehicles. The power required is 2,628 MW or 21.5 M MWh/y. At a cost of $50/MWh, the cost is $1,072 M/y which is easily the major cost in the operation.

The capital cost is $1,970 which includes an allowance of 1% for land and 10% for other offsite facilities on the ISBL cost described above

[18] P. Zubrinich, Solar costs have fallen 82% since 2010. *PV Magazine* June 4, 2020.

Table 4.2: Hydrogen production cost by proton exchange membrane (PEM) electrolysis — central facility.

PRODUCTION	Kt/y	373.94
Power	MW	2628.72
Capital Cost (Capex)	M$	1970.52
ROC (2,10,20)	% Capex	13.57%
	M$/y	267.33
Working Capital	M$/y	0
Operating Cost (Opex)	% Capex	5%
	M$/y	98.53
Water	kt/y	3365.46
Water efficiency	%	95%
Actual Water	kt/y	3542.59
	ML/y	3542.59
Water Cost	$/kL	3
	M$/y	10.6
Power	MW	2628.72
	MWh/y	2.15E + 07
	$/MWh	50
	M$/y	1072.52
By-product Oxygen	kt/y	2991.52
	t/d	8798.59
	$/t	25
	M$/y	74.79
COST OF PRODUCTION	M$/y	1374.21
Hydrogen	$/t	3675.0
	$/GJ (LHV)	30. 64

($1,775M). The return on capital (ROC) for a 2-year construction, a 10% DCF and a 20 y life with no royalty is 13.57% of capital or $267M/y. No account is taken for the cost of stack replacement over the life of the project. The working capital is considered as zero as there is no storage considered for the hydrogen product.

The other operating costs (labour, general maintenance etc.) are set at 5% of the capital cost/year or $98.53M.

The water required for the electrolysis is 3,365 kt/y. If the efficiency of the water use is 95% (i.e. 5% of the water is lost the facility requirement is for 3,542 kt/y. If we take the water density as unity this is 3,542 ML/y. The cost of the water is taken as $3/kL or 10.6M$/y, which assumes water could be sourced from a dedicated plant or a general utility. It is assumed that this cost is sufficient to cover purification to the levels required by the PEM system. The cost of the water is relatively low but could be higher if the water source is brackish or sea water, which would require extensive treatment.

The efficiency of the electrolysis is taken as 58%, which means 42% of the electric power used will appear as heat. That is the cooling demand will be 1,104 MW (this is similar to the cooling demand of a large 2 GW coal-fired power station). This cooling demand would have to be supplied by cooling water or air cooling. The cost of this cooling plant is assumed to be incorporated in the capital cost of the electrolysis plant. However, if cooling water is being used this will increase the overall water demand for the facility which has not been included in this estimate.

Oxygen is produced as a by-product (about 3 Mt/y). This is a large amount of oxygen (about 8,800 t/d). If the oxygen is sold at $25/t, this will produce a by-product credit of $74.8 M/y. This volume of oxygen could satisfy the demand for a large integrated steel operation and this level of demand would not be available in most locations.

Together the total cost for producing 373 kt/y of hydrogen production is $1,374M/y, which generates a unit production cost for the hydrogen is $3,675/t or $30.76/GJ on a lower heating value (LHV) basis. The sensitivity to power cost and received by-product oxygen price is shown in Figure 4.10.

The figure clearly demonstrates the sensitivity of the hydrogen production cost to the power cost. In general, industrial power prices in many urbanised jurisdictions with large industrial operations present, power prices are about $40/MWh. Power prices are often below this in areas where large-scale hydropower is available. Furthermore, for the power demand required for a central hydrogen facility (2,600 MW), such a facility may potentially be a foundation customer for a new major power plant — hydro or nuclear. In this case, taking the analogy of aluminium smelters which are very large consumers of power, it may be feasible to develop a large hydrogen operation with power costs at about $25/MWh or below. Using a power cost of $25/MWh would reduce the estimated hydrogen production costs to

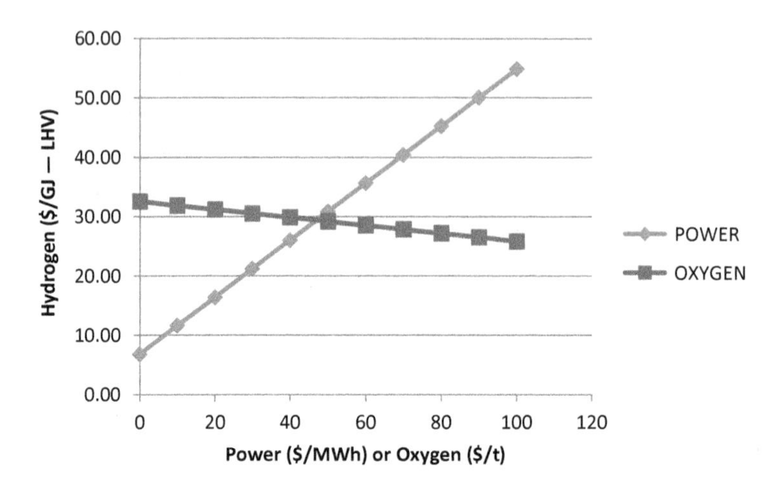

Fig. 4.10: Sensitivity to power cost and oxygen price.

$2,241/t or $18.68/GJ (LHV). Further cost savings could be obtained if the facility employs the cheaper and more efficient alkaline cells.

Also shown in Figure 4.10 is the sensitivity to by-product oxygen price. As noted above, a central facility would produce a very large quantity of oxygen suitable for a large integrated steel mill operation perhaps with other industries such as chemicals and petrochemical operations also being supplied. The value of oxygen to such large operations is typically in the region of $25/t. Sale at this price may reduce the hydrogen production cost by 10%. There may be opportunities for smaller but higher value sales, such as medicinal oxygen, where prices over $100/t can be achieved.

Photochemical Water Splitting

One of the main drivers for hydrogen production is the concept of using sunlight to generate power in a PV system, which is then used to generate hydrogen by water electrolysis. As is indicated by the description above, when taking into account the electricity generating cost this route is likely to prove costly. But not only that, the inefficiencies inherent in the electrolysis of water lead to poor electricity utilization. This would inevitably make hydrogen vehicles less attractive to fully electric battery vehicles as a pathway to decarbonise the transport sector.

One way around this problem is to go directly to the production of hydrogen from water by photochemical splitting. The basis of this method dates back several decades when it was noticed that sunlight impinging on certain colloidal solutions of ruthenium complexes in water induced the splitting of the water molecules into hydrogen and oxygen.[19] Since these early observations there has been a great deal of research in the field, identifying additives and alternative cheaper catalysts for the reaction.[20] The proposed method is illustrated in Figure 4.11.

This method involves a photo-cell comprising a photo-electrode and a counter-electrode. The key is the construction of the photo-electrode. This is a semi-conductor material, typically based on titania doped with active photo-catalysts. The semi-conductor has a valence band and a conduction

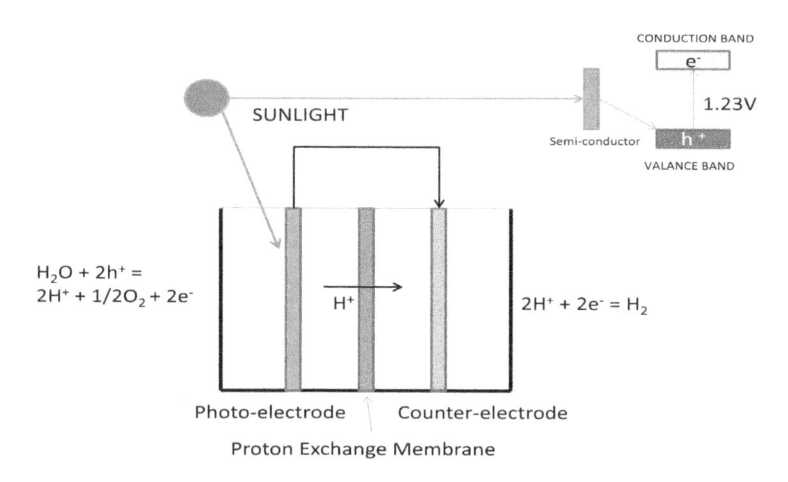

Fig. 4.11: Photochemical water splitting.

[19]Kiwi J, Grätzel M. Hydrogen evolution from water induced by visible light mediated by redox catalysis. *Nature* **281**, 657–658 (1979) and Kiwi J, Grätzel M. Projection, size factors, and reaction dynamics of colloidal redox catalysts mediating light induced hydrogen evolution from water. *J Am Chem Soc* 1979 **101** (24), 7214–7217.

[20]Yu J, He Q, Yang G, Zhou W, Shao Z, Ni M. Recent Advances and Prospective in Ruthenium-Based Materials for Electrochemical Water Splitting. *ACS Catal.* **2019** *9* (11), 9973–10011. N. Fajrina and M. Tahir "A critical review in strategies to imp[rove photocatalytic water splitting towards hydrogen production", *J. Hydrogen Energy*, **44** (2) 540–577 (2019) and S. J. Mun and S-J Park, "Graphitic Carbon Nitride materials for Photocatalytic Hydrogen production via Water Splitting: A Short Review", Catalysis, 9, 805 (2019)

band separated by at least 1.23 V, which is the energy requirement for splitting water (as for electrolysis). When photons of suitable energy impinge on the semi-conductor an electron is ejected from the valence band into the conduction band and generating an electron hole in the valence band. In theory this could be affected by photons in the near infrared but in practice higher energy photons are required to overcome inefficiencies, particularly in the photo-electrode (c.f. electrolysis anode over-potential).

The hole in the valence band is then quenched by a water molecule to form oxygen and two protons. This is a simplified form of the reaction which is more complex and differs in alternative media (alkaline or acid conditions). The protons pass through a PEM to the counter-cathode where hydrogen is generated. This is a relatively simple process and the counter-electrode can be formed from platinised graphite.

Photochemical water splitting is still in the early stages of development, it is not possible at this stage to give an estimate of hydrogen production costs.

Hydrogen Production from Biomass

An alternative approach to producing carbon emission free hydrogen is to use biomass as a feedstock. The term biomass covers agricultural products, which can be produced sustainably. Preferably the crop is produced without the recourse to industrial fertilizers which produce carbon emissions in their production, e.g. forest wood waste, prairie grasses and the like. In some cases the biomass is agricultural by-products of a different crop such as corn stover (maize stalks) or bagasse (sugar cane waste). In this instance the carbon emissions from fertilizer due to the production of the principal crop are associated with that crop and consequently carbon emission from growing the biomass by-product is considered to be zero.

To confuse matters somewhat, certain streams in municipal waste collection are regarded as sources of renewable fuel and also included in the term biomass.

The basic method for producing hydrogen from these diverse materials is by gasification (similar to coal gasification which is described in the above Chapter 2). The process involves gasification to form synthesis gas, gas clean-up, water-gas-shift to maximise hydrogen content and then separation of the hydrogen.

Biomass Logistics

The principal problem with biomass gasification is the nature of the feedstock. Raw (as received) biomass has very low specific energy, typically only 5–6 GJ/t. This is a consequence of the inherently high water (moisture) content of material as received at the facility. Drying of the 'as received' biomass will lift the specific energy to about 12 GJ/t, but this is hardly sufficient to maintain the smooth operation of a gasifier and in many operations a supplementary fuel is added to raise the fuel specific energy higher. Adding natural gas is an obvious choice and this is practised in many large facilities for the gasification or incineration of urban waste; however, this defeats the object of producing a carbon emission free product.

Furthermore, the molecules that compose biomass contain a lot of oxygen. Biomass is generally based on a poly-cellulose matter, which has an oxygen content of 55%. This bound oxygen cannot be removed prior to gasification.

One practical problem is storage. Biomass, as received or dry, can undergo putrefaction and spontaneous combustion brought about by bacteria and the like in the biomass (compost heap combustion is familiar to most gardeners and horticulturalists), the prevention of this increases costs.

Another issue is the variability of the biomass quality (specific energy), even dedicated biomass varies according to season. This variability can influence the downstream performance of the gasifier. The variability is not due to water content alone but also the nature of the material making up the biomass varies according to season. In addition dedicated biomass crops (prairie grass, forest waste) can be exposed to total loss through wild-fires and the like.

Many studies for bio-gasification assume that the biomass is available for free. This is not so. The use of biomass will inevitably lead to loss of soil carbon and essential minerals necessary for healthy growth. If the biomass is produced as a by-product to a cash crop, then the farmer (or forester) would want some form of compensation for this.

Finally the collection costs of transporting the biomass to a gasifier operation will be significant (typically \$5–\$15/t). The logistics of transporting the biomass will incur some carbon emissions from fuel. The cost and transport logistics will place limits to the scale of a biomass gasifier. As the distance from the biomass source rises so does the cost of biomass transport. This compromise limits the scale of bio-gasifiers in comparison to the large-scale gasification operations for coal. That is the full economy of scale is not realised.

Biomass Quality

Zhou *et al.*[21] have compiled proximate and ultimate[22] analysis of materials constituting municipal waste, some of which are presented in Table 4.3 along with the analysis for corn stover and prairie grass. These are compared to lignite and a typical black coal used for thermal power generation.

Table 4.3: Analysis for some biomass materials of interest.

	Specific energy (GJ/t)	**Water%**	**Ash%**	**C%**	**H%**	**O%**	**N%**	**S%**	
Vegetable	16.8			44.9	5.5	45.4	3.6	0.6	DAF[a]
Bone	15.7			58.0	7.2	25.4	8.7	0.74	DAF
Paper	15.1			45.5	6.3	47.7	0.2	0.2	DAF
Dry wood	19.6			50.5	5.9	43.4	0.11	0.03	Pine — DAF
Wood	12.6	28.6	28.6	37.8	4.5	28.4	0.07	0.0	As received
Corn stover	12.6	20.0	6.0	35.0	4.4	34.1	0.5	0.01	As received
Prairie grass	14.6	15.0	6.2	39.5	4.8	33.8	0.7	0.08	As received
Lignite	15.6	64.7	7.4	18.7	1.3	7.7	0.14	0.07	Victoria
Black coal	25.3	8.7	13.5	60.6	3.8	11.6	1.33	0.44	Queensland

[a] Dry Ash Free.

[21] Zhou H, Meng A, Long Y, Li Q, Zhang Y. (2014). Classification and comparison of municipal solid waste based on thermochemical characteristics. *J Air Waste Manag Assoc* **6:** 597–616.

[22] Proximate analysis and ultimate analysis are standard analytical procedures for combustion fuels.

As received, municipal waste has a specific energy of typically 5–6 GJ/t. After drying, the remaining vegetable matter, bone and paper have specific energies of about 15–17 GJ/t. Dry wood has a specific energy of nearly 20 GJ/t. On a dry ash-free basis (DAF), these materials typically have 55% carbon plus hydrogen. The exception is bone matter which has over 65% carbon plus hydrogen. The main other element is oxygen which can be over 40% of the total. Vegetable matter and bone have relatively high levels of nitrogen and sulphur.

As received, biomass such as wood, corn stover, prairie grass has high moisture content. This encourages bacterial attack in storage and degrades the efficiency of the gasifier. For optimum performance, this moisture is removed prior to storage and gasification, typically by steam driers using excess steam produced in the gasification process.

The moisture content of as received biomass is generally lower than lignite (brown coal), a widely used fossil fuel for power generation and gasification. Lignite from the La Trobe Valley in Victoria, Australia, has moisture content over 60% and is used as a power generation fuel. Lignite used to fuel the Dakota Coal Gasification Company operation has a moisture content of about 25%–40%. However, lignite has lower fixed oxygen content than biomass giving them a higher calorific value measured on a DAF basis.

Compared to biomass and lignite, black coal has a much lower as-received moisture content and lower fixed oxygen, which gives it a high specific energy. The value in the table is for a typical export quality thermal coal from Queensland, Australia.

Process Flow Sheet for Biomass Gasification

The process flow-sheet to produce hydrogen from biomass is similar to that for coal gasification, Figure 4.12.

The gasifier island (shown in more detail in Figure 4.13) converts the biomass into synthesis gas (carbon oxides and hydrogen). The high-temperature water–gas shift (WGS) process converts carbon monoxide into more hydrogen. The dry gas exiting this unit operation is substantially carbon dioxide and hydrogen. This is passed to the acid gas plant which strips out the carbon dioxide and hydrogen sulphide making them

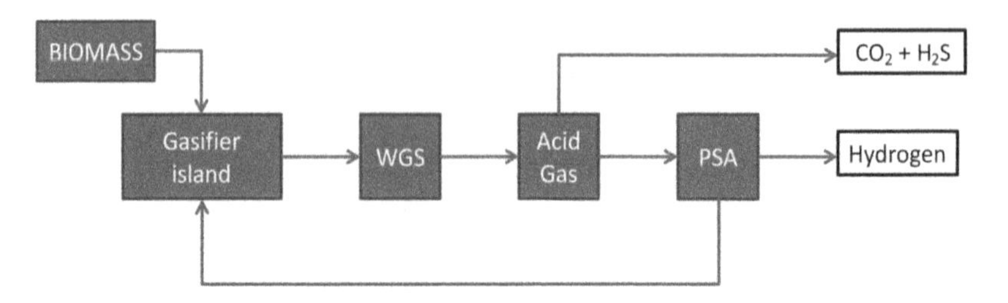

Fig. 4.12: Process flow for biomass to hydrogen.

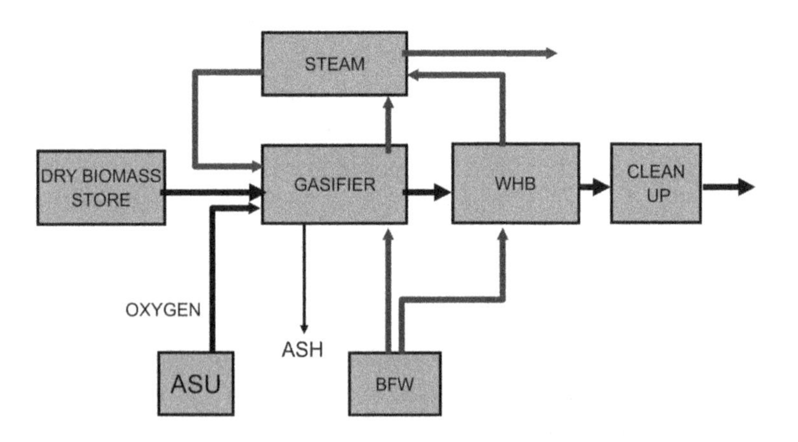

Fig. 4.13: Gasifier island for biomass gasification.

available for geo-sequestration or sulphur recovery. Because biomass is considered as a renewable fuel, the carbon dioxide would not necessarily require disposal by geo-sequestration or carbon credit purchase. However, many approaches embrace mixing the biomass with a fossil fuel in a co-firing operation in which case, for zero carbon emission, a disposal mechanism for some or all of the carbon dioxide would be required. The hydrogen rich gas exiting the acid gas plant would pass through a pressure swing adsorption (PSA; or membrane) separation system to produce the pure hydrogen product. Tail gas from the hydrogen separator would be recycled to the gasifier with a small quantity flared to extract inert gases (nitrogen, argon) and prevent build up in the recycle loop.

The gasifier island (Figure 4.13) comprises several unit operations. It may be best practice that the biomass is pre-dried and stored with an inert gas environment. This may be avoided if storage is minimal and the gasifier is of a type that it can accomplish the drying duty.

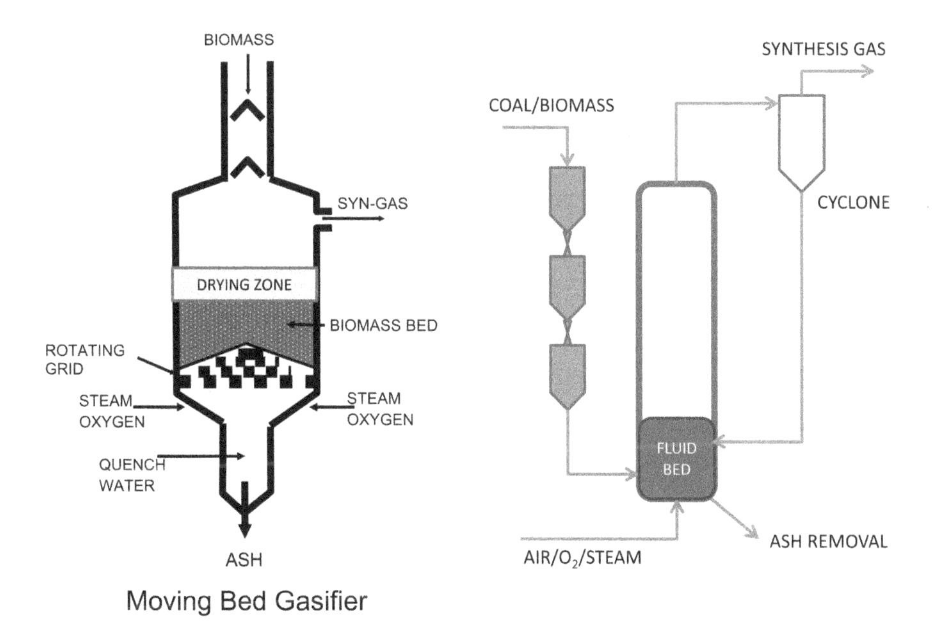

Fig. 4.14: Gasifiers for biomass feedstock.

There have been many approaches to the design of gasifiers for biomass.[23] The properties of biomass (high water content, high fixed oxygen content and suitable scale of operation) reduce the selection of the types available. The gasifier designs based on moving-bed gasifier[24] and the fluid-bed gasifier,[25] which are described in Chapter 2 for coal gasification, have been most popular for biomass — they are relatively small and well proven for this type of feedstock (Figure 4.14).

In the moving-bed gasifier, the design can accommodate a drying section above the gasifier bed. This performs the duty of removing moisture not removed by biomass pre-treatment prior to gasification. Fluid-bed type has been proposed which require milling and sizing of the biomass.

---

[23] Craig KR, Mann MK, (October 1996) *Cost and Performance Analysis of Three Integrated Biomass Gasification Combined Cycle Power Systems*. NREL. US DOE DE-AL36-83CH10093.

[24] Beenackers AACM. (1999), Biomass gasification in moving beds, a review of European technologies. *Renewable Energy*, **16** (1–4): 1180–1186.

[25] van den Enden PJ, Lora ES. (2004) Design approach for a biomass fed fluidized gasifier using the simulation software CSFB. *Biomass and Bio Energy* **26** (3): 281–287.

In this case to avoid excessive burning of feedstock in the gasifier to generate the heat required to remove the moisture, this type of gasifier benefits from pre-drying the raw biomass. This can be done by using steam generated by the gasifier.

One of the major problems with biomass gasification is that the fixed oxygen content and consequential low specific energy results in a relatively low temperature operation, even with pure oxygen rather than air as the oxidant supplied by an air separation unit (ASU). This produces excessive tars and pyrolysis liquids in the synthesis gas so that clean-up cannot be simply achieved with candle filter systems. Washing the raw gas with a solvent (such as methanol) is widely recommended. One of these processes (Rectisol^{TM}) can handle the full range of impurities found in the pyrolysis gases and can be used to extract carbon dioxide, unfortunately it is capital intensive and operationally complex.

Economics of Biomass to Hydrogen

The conversion of biomass to transport fuels by the Fischer–Tropsch process has been extensively studied, these studies lay out the main issues for the conversion of biomass into hydrogen. The production of liquid products by the Fischer–Tropsch process requires biomass gasification to synthesis gas, water-gas-shift followed by the Fischer–Tropsch conversion to liquids. Production of hydrogen should require less capital as it avoids the Fischer–Tropsch conversion step and the refining of the liquid products into transport fuels. These studies also address the issue of biomass collection and logistics and hence limits to the size of the facility.

The following economic descriptive has been adapted from a study by Kreutz *et al.*[26] for the conversion of 1 Mt/y of corn stover into Fischer–Tropsch liquids. A proximate and ultimate analysis for the corn stover is given in Table 4.4. On a dry basis the stover had a specific energy of 18.7 GJ/t (Higher Heating Value, HHV).

Feedstock and product flow rates are given in Table 4.5. The hydrogen production has been estimated assuming the efficiency of the extraction

---

[26] Kreutz TG, Larson ED, Lui G, Williams RH. (2008). Fischer–Tropsch Fuels from Coal and Biomass. *25th Annual International Pittsburgh Coal Conference.*

Table 4.4: Proximate and ultimate analysis of the corn stover feedstock.

Corn Stover	Mt/y	1
Proximate	as received	
Fixed C	wt%	18.10%
Volatile matter	wt%	61.60%
Ash	wt%	5.30%
Moisture	wt%	15.00%
Total		100.00%
LHV	MJ/kg	14.509
HHV	MJ/kg	15.935
Ultimate	dry basis	
Carbon	wt%	46.96%
Hydrogen	wt%	5.72%
Oxygen	wt%	40.18%
Nitrogen	wt%	0.86%
Sulphur	wt%	0.09%
Ash	wt%	6.19%
Total		100.00%
HHV	MJ/kg	18.748

Table 4.5: Feedstock and product flow rates.

Stover as received	t/d	3581
dry	t/d	3044
DAF	t/d	2822
	MW LHV	601
	MW HHV	660
Oxygen	t/d	59
Purity	mass%	99.54%
Oxygen for gasifier	t/d	58
Hydrogen	kmol/s	1.15
	kg/s	2.31
	t/y	67,790.87
	kt/y	67.79
	t/d	199.4

Table 4.6: Power balance for biomass to hydrogen facility.

Production	MW	65.99
Use		
Biomass	MW	0.41
Lock hopper	MW	0.35
Synthesis gas clean-up	MW	3.86
Fuel gas comp	MW	0.08
Steam cycle	MW	1.27
ASU	MW	14.78
O_2 Compressor	MW	6.55
Total consumption	MW	27.3

by PSA is 85%. The result of the simulation is that 1 Mt/y of corn stover is converted into about 68,000 t/y hydrogen.

A power balance for the system is given in Table 4.6. The steam and power system is capable of producing about 66 MW of power of which 27 MW is used within the facility. The use is dominated by the power demand of the ASU and the attached oxygen compressor (total 21 MW). The synthesis gas clean-up (Rectisol™) consumes about 4 MW.

There is an excess of power generation over the demand of nearly 39 MW, which is exported from the facility.

The capital cost breakdown is given in Table 4.7. The cost estimates for 2007 have been escalated to 2018 values in the final column. An estimate for a PSA unit for the extraction of the hydrogen product has replaced the cost of the Fischer–Tropsch conversion. The capital cost estimate is \$916M for 2018 construction.

Using the above statistics for a biomass to hydrogen facility, the estimate for the hydrogen production cost is developed in Table 4.8. It is assumed that the facility will take 3 years to build and require a 10% discounted cash flow return, operate for 20 years and pay a 2% capitalised royalty will require an ROC of 14.6%. Working capital is assumed to be zero with no hydrogen storage.

Non-feedstock operating costs are \$59.6 million/year. The feedstock (1 Mt/y) is assumed to cost \$10/t to cover collection and delivery costs.

Table 4.7: Capital cost breakdown for a biomass to hydrogen facility.

		2007		2018
Land	$M	$5.00	0.74%	$6.78
Gasifier	$M	$266.00	39.35%	$360.81
Rectisol and water–gas shift (WGS)	$M	$58.00	8.58%	$78.67
ASU	$M	$94.00	13.91%	$127.50
Pressure swing adsorption (PSA)	$M	$10.00	1.48%	$13.56
Power Island	$M	$64.00	9.47%	$86.81
		$497.00	73.53%	$674.14
Engineering	+15%	$74.55	11.03%	$101.12
Contingency	+10%	$49.70	7.35%	$67.41
Start up, spares	+6%	$29.82	4.41%	$40.45
Off-sites	+5%	$24.85	3.68%	$33.71
TOTAL	$M	$675.92	100%	$916.83

Table 4.8: Estimate for the cost of hydrogen production from biomass.

2018 Capital cost (CAPEX)		$M	$916.83
ROC (3-year build, 10% DCF, 20-year life)		% CAPEX	14.60%
		$M/y	$133.88
Working Cap		$M/y	none
Operating cost	% CAPEX		
Labour	1%	$M/y	$9.17
Maintenance	3.00%	$M/y	$27.51
Insurance	1.50%	$M/y	$13.75
Catalysts and chemicals	1%	$M/y	$9.17
Subtotal			$59.59
Total Fixed costs		$M/y	$193.47
Feedstock		Mt/y	$1.00

(*Continued*)

Table 4.8: (*Continued*)

Feedstock costs	$10/t	M$/y	$10.00
Total annual costs		$M/y	$203.47
By-product electricity		MW	38.69
		MWh/y	315,710.4
		$/MWh	50
		$M/y	$15.79
NET Cost		$M/y	$187.69
Hydrogen unit cost		**$/t**	**$ 2,768.63**
		$/GJ	$23.08

The total annual costs of $203.5 million are reduced by $15.8 million/year from the sale of by-product electricity at $50/MWh. The net production costs are $187.7 million/year. This generates a hydrogen production cost of $2,768/tonne or $23.08/GJ based on the LHV.

Review

It is useful at this point to review the cost estimates for hydrogen production, which have been described in the Chapter 3 and Chapter 4. These are summarised in Tables 4.9 and 4.10. For the fossil fuel cases, these tables incorporate the quantity of carbon dioxide produced and the cost of carbon dioxide disposal at $50/t carbon dioxide.

Table 4.9 summarises the base case data as presented in the Chapter 3 and Chapter 4. The table gives the cost by gasification methods which can be compared to the estimate for electrolysis with power at $50/MWh at $3675/t; the estimated cost of hydrogen by electrolysis with power at $25/MWh is $2,240/t.

Table 4.10 summarises the estimates when the cost of the gas feedstock for the natural gas cases reflects more realistically real-world costs. That is lower scale operations face higher feedstock costs. Since electric power can be distributed over a wide area at more or less the same cost and the low economy of scale for electrolysis operations, the cost of

Table 4.9: Summary of base case hydrogen cost estimates.

		H_2 micro GAS	H_2 small GAS	Industrial scale — Gas	Large gas facility	ICI gas	NAS gas	COAL PSA	COAL methanation	Biomass
		Figure 3.4	**Figure 3.4**	**Figure 3.4**	**Figure 3.4**	**Table 3.3**	**Table 3.3**	**Table 3.5**	**Table 3.5**	**Table 4.8**
Production	kt/y	1.76	26.20	150.00	300.00	373.94	373.94	1042.72	838.90	67.79
Capex	$M	10.25	85.67	337.69	582.27	692.35	532.37	3321.24	3321.24	916.83
Construction period	years	3	3	3	3	3	3	4	4	3
Plant life	years	15	15	15	15	20	20	20	20	20
Return on capital	%/y	16.34%	16.34%	16.34%	16.34%	14.60%	14.60%	15.15%	15.15%	14.60%
Non-gas operating costs	$M/y	2.34	19.57	77.14	133.02	146.10	112.34	847.49	847.49	193.47
Feedstock usage	PJ/y	0.40	5.94	34.00	68.00	84.76	84.76	244.79	244.79	17.67
Feedstock cost	$/GJ	2.00	2.00	2.00	2.00	2.00	2.00	1.00	1.00	1.00
Feedstock costs	$M/y	0.80	11.88	68.00	136.00	169.52	169.52	244.79	244.79	10.00
By-product credits	$M/y	0.04	4.44	25.41	50.83	63.35	63.35	3.02	3.02	15.79
Greenhouse gas (GHG) emissions	$MtCO_2e/y$	0.020	0.297	1.700	3.400	4.238	4.238	20.084	20.084	0
GHG emission cost ($50/$tCO_2$)	$M/y	1.00	14.84	85.00	170.00	211.90	211.90	1004.22	1004.22	0
Production costs with GHG disposal	$M/y	4.10	41.85	204.73	388.19	464.17	430.41	2093.48	2093.48	187.69
Hydrogen costs	$/t	2328.53	1597.69	1364.86	1293.97	1241.29	1151.01	2007.70	2495.51	2768.63
(Non-GHG cost)	$/t	1761.87	1031.02	798.20	727.30	674.62	584.34	1044.63	1298.44	2768.63

The estimated cost of hydrogen by electrolysis with power at $50/MWh is $3,675/t (Table 4.2)
The estimated cost of hydrogen by electrolysis with power at $25/MWh is $2,240/t (Figure 4.10)

Table 4.10: Estimates costs as per Table 4.9 with variation in feedstock price.

		H_2 micro GAS	H_2 small GAS	Industrial scale — gas	Large gas facility	ICI gas	NAS gas	COAL PSA	COAL methanation	Biomass
		Figure 3.4	**Figure 3.4**	**Figure 3.4**	**Figure 3.4**	**Table 3.3**	**Table 3.3**	**Table 3.5**	**Table 3.5**	**Table 4.8**
Production	kt/y	1.76	26.20	150.00	300.00	373.94	373.94	1042.72	838.90	67.79
Capex	$M	10.25	85.67	337.69	582.27	692.35	532.37	3321.24	3321.24	916.83
Construction period	years	3	3	3	3	3	3	4	4	3
Plant life	years	15	15	15	15	20	20	20	20	20
Return on capital (ROC)	%/y	16.34%	16.34%	16.34%	16.34%	14.60%	14.60%	15.15%	15.15%	14.60%
Non-gas operating costs	$M/y	2.34	19.57	77.14	133.02	146.10	99.94	847.49	847.49	193.47
Feedstock usage	PJ/y	0.36	5.35	30.64	61.28	76.39	65.52	180.95	180.95	17.67
Feedstock cost	**$/GJ**	**12.00**	**8.00**	**4.00**	**3.00**	**3.00**	**3.00**	**1.50**	**1.50**	**1.00**
Feedstock costs	$M/y	4.31	42.81	122.56	183.85	229.16	196.56	271.42	271.42	10.00
By-product credits	$M/y	0.02	2.66	15.25	30.50	38.01	32.61	3.02	3.02	15.79
Greenhouse gas (GHG) emissions	$MtCO_2e/y$	0.018	0.268	1.532	3.064	3.819	3.276	20.084	20.084	0
GHG emission cost ($50/t$CO_2$)	$M/y	0.90	13.38	76.60	153.20	190.96	163.80	1004.22	1004.22	0
Production costs with GHG disposal	$M/y	7.53	73.10	261.06	439.57	528.21	427.70	2120.11	2120.11	187.69
Hydrogen costs	**$/t**	**4279.75**	**2790.33**	**1740.41**	**1465.24**	**1412.56**	**1143.78**	**2033.25**	**2527.26**	**2768.63**
(Non-GHG cost)	$/t	3769.07	2279.64	1229.72	954.55	901.88	705.73	1070.17	1330.19	2768.63

The estimated cost of hydrogen by electrolysis with power at $50/MWh is $3,675/t (Table 4.2)

The estimated cost of hydrogen by electrolysis with power at $25/MWh is $2,240/t (Figure 4.10)

hydrogen by electrolysis is not considered to be materially different from the base case in Table 4.9.

Small-Scale Hydrogen Production

The first two columns in the tables give estimates for the production of low volumes of hydrogen by gasification/steam reforming of natural gas. The interest is in the comparison with electrolysis produced hydrogen. Table 4.9 shows that electrolysis is a poor option for the base case, however, at more realistic gas prices for smaller users (Table 4.10) electrolysis would be a preferred option for very small volumes (1.76 kt/y) of hydrogen produced on a carbon free basis and is competitive with the gas route at the small-scale operations (26.2 kt/y).

Larger Scale Hydrogen Production

The next four columns describe the hydrogen cost of production from natural gas at increasing larger scales of production (from 150 kt/y typical of large-scale industrial production to the 373 kt/y required for a central hydrogen distribution facility) by the gasification/steam reforming route. The first three columns in this group are built on data using the now largely superseded methanation method for hydrogen purification. The National Academy of Sciences (NAS) data is probably more realistic with lower capital as a consequence of elimination of low-temperature water-gas-shift and the use of a PSA separator.

Gasification/steam reforming methods gain competitive advantage over electrolysis at larger scales of production. As well as gaining through economy of scale, the gasification route also gains advantage of being able to access lower gas prices in most jurisdictions ($3/GJ). The electrolysis route may have a cost advantage in regions of inherently high gas prices, such as regions relying on LNG for gas supply but is especially relevant if those locations have access to low-power costs.

Hydrogen from Coal and Biomass

The next two columns summarises the cost estimates from coal and the final column from biomass. Coal routes are high in capital cost and as a consequence would be built for very large production of hydrogen to get a reasonable economy of scale benefit. The two cases are for a PSA route and a methanation system for hydrogen purification.

These estimates suggest that electrolysis in not competitive with large-scale coal production on a carbon free basis using the more efficient PSA route. However, given the inherent errors in these estimates, at low power costs the electrolysis route is competitive with the methanation route. A deduction is that for coal-based routes it is critically important to develop an optimum design so as to minimise the capital and operating costs.

The biomass route for hydrogen production, although nominally giving carbon free hydrogen, is very similar to the costs of producing hydrogen by electrolysis.

CHAPTER 5

HYDROGEN STORAGE AND TRANSPORT

The Chapter 4 discussed the methods of production and cost of producing hydrogen from various sources. Production is only the first part of the problem for a hydrogen-based economy. Of equal significance is the method and costs of hydrogen storage and transport. This chapter and the Chapter 6 will review the main methods and issues relating to hydrogen storage and transport from production facility to the end user.

Included in this discussion are the methods and cost of compression. For the most part hydrogen is produced at a low pressure when compared to the preferred delivery pressure for vehicles at 40 MPa–80 MPa (400–800 bar). The cost of compression of hydrogen from near atmospheric pressure to that required is significant in terms of both capital and the energy required.

Hydrogen Storage in Cylinders

We are all familiar with the storage of hydrogen in cylinders, reference to these gives an idea of the scope of the problem with storage and transport. Some common cylinders are described in Table 5.1.

Table 5.1: Common cylinders for transporting hydrogen.

		Steel cylinders				
		BOC[(a)] K	BOC L	Composite		Coselle
Length	cm	146	1640	261.6	136	100,000
Diameter	cm	23	23	43.5	46	23
Pressure	bar	150	230	350	700	230
Weight	kg	65	87	178	140	5305
Volume	cc	60,667	681,468	270,000	130,000	41,552,950
Hydrogen	L	60.7	681.5	270.0	130	41,553.0
Hydrogen	kg	0.81	13.99	6.20	5	853.32

[(a)] BOC (British Oxygen Company) is a subsidiary of Linde AG.

Two types of steel cylinder are described (BOC K, BOC L[1]). Typical small cylinders are the BOC K type. They are about 150 cm in length and 23 cm in diameter and are capable of holding hydrogen at 150 bar (15 MPa). The cylinder weighs 65 kg. The volume of hydrogen held is 60.7 L that will weigh only 0.81 kg.

Considering that for a practical use in a vehicle we would require about 3 kg of hydrogen for an average weekly use, we would require four such cylinders weighing 226 kg. The larger cylinder (BOC L type) will deliver more hydrogen, 14 kg of hydrogen at 230 bar (23 MPa), but is over 16 m long. This type is used to deliver hydrogen by trailer truck but completely unsuitable for use in a vehicle in its own right.

This problem of vehicle on-board hydrogen storage has largely been solved by the move to composite cylinders.[2] These are fabricated from multiple layers comprising a hydrogen barrier that is either aluminium or polymer surrounded by carbon-fibre reinforced epoxy and are further strengthened by wrapping the cylinder in high-tensile steel wire or similar filament.

---

[1] BOC website: https://www.boconline.co.uk/en/health-and-safety/gas-safety/cylinder-weights-sizes/cylinder-weights-size/index.html

[2] Sirosh N. (2002) Hydrogen Composite Tank Program. In: *Procs. 2002 U.S. DOE Hydrogen Prog Rev NREL/CP-610-32405*; and see also Steelhead Composites: https://steelheadcomposites.com/wp-content/uploads/2018/04/Hydrogen-Brochure.pdf

Two examples of composite tanks are given, one of 2.61 m × 0.43 m holding 270 L of hydrogen at 350 bar (35 MPa) holding over 6 kg of hydrogen and one suitable for most vehicle operations at 1.36 m × 0.46 m holding 130 L (5 kg) of hydrogen at 700 bar (70 MPa). These results go a long way towards solving the issue of providing fuel for passenger vehicles and small commercial vehicles, multiple units could be used for the movement of trucks.[3] However, it is not satisfactory for the movement of large volumes of hydrogen such as that required for a central hydrogen distribution facility.

One solution for mass transport has been proposed in the 'coselle' concept. This method is being developed primarily for natural gas transport but could be applied to hydrogen. In this method a 10-km long, 23-cm diameter steel pipe is coiled into a coselle. The coselle takes gas at 230 bar that is equivalent to each coselle carrying 850 kg of hydrogen. This system is discussed further in the next chapter.

Transport of Hydrogen as Liquid

For the mass transport of hydrogen, if transport as compressed gas is a problem, then transport as liquid hydrogen can be considered. The contraction ratio is 848:1 and this will clearly reduce the volume of material to be transported. The problems with this solution are that the boiling point of hydrogen is −253°C (20 K) so that transport has to be by a cryogenic container and the energy required to liquefy hydrogen gas is high.

Hydrogen Liquefaction

In conventional liquefaction, such as the liquefaction of nitrogen, the gas is compressed to about 20 MPa, cooled and then expanded in a Joule-Thompson (J-T) expansion valve. The work done in the expansion cools the gas that passes through the heat exchanger and back to the compressor. Repeating the cycle cools the gas further until eventually the gas liquefies.[4] This cannot be applied to hydrogen because the hydrogen has

[3] In fact composite tanks for CNG storage is now well established for buses and trucks.

[4] Timmerhaus C, Flynn TM. (1989) *Cryogenic Engineering*. Plenum Press, New York, NY.

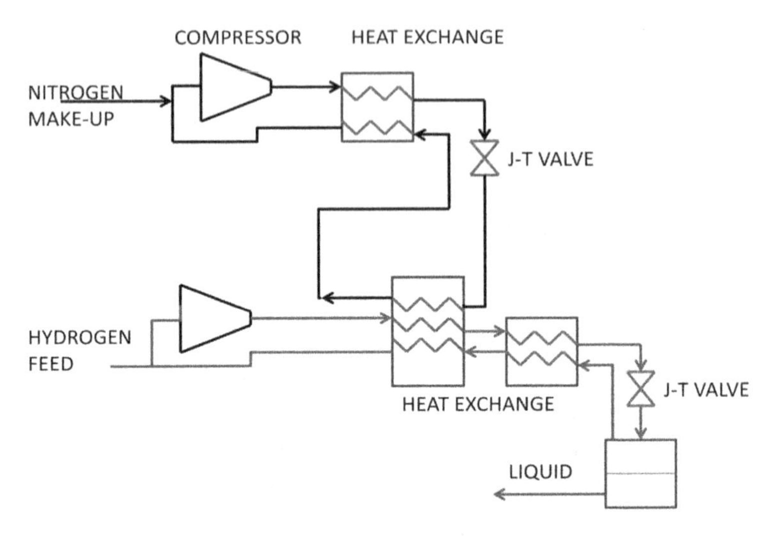

Fig. 5.1: Process outline for producing liquid hydrogen.

to be cooled to below the J-T inversion temperature (−72°C, 201 K) before the gas cools on an expansion cycle. This is commonly done by coupling a nitrogen liquefaction cycle into the process to cool the hydrogen below the J-T inversion temperature as is illustrated in Figure 5.1. There are several variants on the process such as incorporating more heat exchange or a liquid nitrogen bath for the heat exchange.

Coupling the nitrogen liquefaction cycle with the hydrogen liquefaction cycle clearly increases the capital costs. As will be discussed later, the hydrogen compressors are different from those for nitrogen and are more costly to install and maintain. Furthermore, the energy required will be the sum of the nitrogen liquefaction (ideal 0.207 kWh/kg) and the hydrogen liquefaction (ideal 3.22 kWh/kg). Note the much higher energy requirement for hydrogen liquefaction.

Ortho and para hydrogen

A complicating feature for hydrogen is that it exists in two forms — *ortho* and *para*. Each atom of hydrogen comprises a proton and an electron and the proton has a spin (value 1/2). For a hydrogen molecule (di-hydrogen) comprising two hydrogen atoms, if the spins of the atoms are aligned, the hydrogen is in the *ortho*-state and if they are opposed, the molecule is in the *para*-state.

These two forms can spontaneously interconvert. At room temperature the thermodynamic equilibrium favours the *ortho*-state and the composition is typically 75% *ortho*- and 25% *para*-hydrogen. At liquid hydrogen temperatures (20 K) the thermodynamic equilibrium strongly favours the *para*-form (nearly 100%). Rapid cooling and liquefaction of the hydrogen 'freezes' the liquid hydrogen dominantly in the *ortho* form. This means the liquid is in a meta-stable state and will slowly re-equilibrate to become dominantly *para*-hydrogen.

Conversion of *ortho*-hydrogen into *para*-hydrogen proceeds with the release of 527 kJ/kg of heat. As hydrogen converts from the *ortho*-dominant room temperature to form *para*-hydrogen favoured for the liquid this heat will be released. This is a particular issue with liquid hydrogen storage where the slow release of this heat increases the boil-off rate and leads to high hydrogen losses.

The solution to the problem is to incorporate a catalyst in the cold side of the hydrogen liquefaction process. The catalyst is typically iron supported on carbon and this can be used as a coating on the insides of the cold heat exchanger operating at about 77 K. The liquid nitrogen heat exchange removes the reaction heat. This converts most of the *ortho*- into *para*- and significantly reduces spontaneous heating on storage.

Storage of Liquid Hydrogen

Because of its low temperature, liquid hydrogen is stored in cryogenic vessels. Spherical vessels have the lowest surface area and are preferred for large quantities of storage. NASA has built a large tank that can contain 225,000 kg of hydrogen (see Figure 5.2).

There are several variants with general construction principles illustrated in Figure 5.3.

The spherical tank comprises several concentric layers. There is an outer insulation layer that may encase liquid nitrogen in the annulus. An inner layer comprises aluminised Mylar or Perlite and in some cases a

Fig. 5.2: Large hydrogen storage tank (NASA[5]).

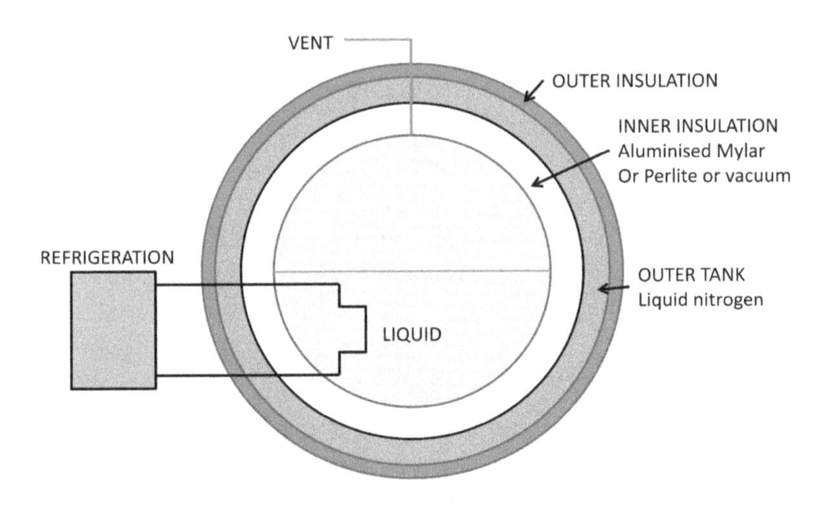

Fig. 5.3: Liquid hydrogen tank construction.

vacuum. The hydrogen liquid in the inner tank is allowed to slowly boil off through a vent-type system. For large storage and to prevent excessive boil-off (due for example from *ortho*-hydrogen to *para*-hydrogen[5]

[5]NASA: https://www.nasa.gov/content/liquid-hydrogen-the-fuel-of-choice-for-space-exploration

Fig. 5.4: Typical cylindrical tank for liquid hydrogen storage (photo courtesy of Linde AG).

conversion), the lower part of the vessel is equipped with a refrigeration system run from plant external to the storage sphere.

For the storage of smaller volumes of liquid hydrogen, e.g. for vehicle filling operations, storage vessels of a cylindrical construction are used, again with multiple insulation layers; Figure 5.4.

Clearly the storage of hydrogen in cryogenic storage tanks comes at high capital cost and an increased operating cost from hydrogen boil-off.

Storage in LNG-type tanks

There are three types of storage tanks in use for the mass storing of liquefied natural gas at −165°C. These are the older single containment type, which are preferred in Japan and the United States, the double containment tank and the full containment tank preferred in Europe.

At the time of writing, it is not clear if LNG storage tanks are suitable for liquid hydrogen storage or if they can be modified for this role, e.g. by adding further insulation and internal refrigeration coils.

Storage in Salt Caverns

In order to avoid the high cost of liquid hydrogen storage, very large volumes of gas can be stored in salt caverns, provided the local geology is suitable for their construction. Salt caverns for gas storage have a long history. They were patented in 1916 and have been widely used in the chemicals industry since the 1950s. They offer absolute tightness, i.e. no leaks and very low losses. Construction is conducted entirely from the surface in several stages and is cheap to perform. They are well known for storing ethylene, LPG and hydrogen; Hevin[6] gives a good overview of the subject.

Salt (sodium chloride, halite) deposits are widespread throughout the world. Many of the deposits are very large and are often found in the sedimentary basins associated with oil and gas. The deposits occur in several forms such as thick beds and salt dome intrusions. Salt caverns are made by solution mining in the following phases:

- The salt layer of suitable thickness and depth (about 1 km below the surface) is drilled.
- Water is pumped to the bottom of the well which dissolves the salt. The brine is brought to the surface via the annulus of the drill for disposal. Disposal can be by evaporation ponds or passing to a chlor-alkali works for producing chlorine and sodium hydroxide.
- The drill is partially withdrawn and leaching continues developing the cavern in an upwards direction. Insoluble matter in the salt falls to the base of the cavern. Sonar methods are used to control the shape and volume of the cavern.
- When complete, the gas to be stored can be pumped into the cavern through the annulus and brine withdrawn or added from an inner drill pipe from the base of the cavern.

The operating method is illustrated in Figure 5.5.

The maximum operating pressure of the cavern is determined by the depth of the cavern from the surface. Hydrogen gas is pumped into the top

---

[6]Hevin G. (2019) Underground storage of Hydrogen in salt caverns. In: *European Workshop on Underground Energy Storage*, Paris, France (7–8 November 2019).

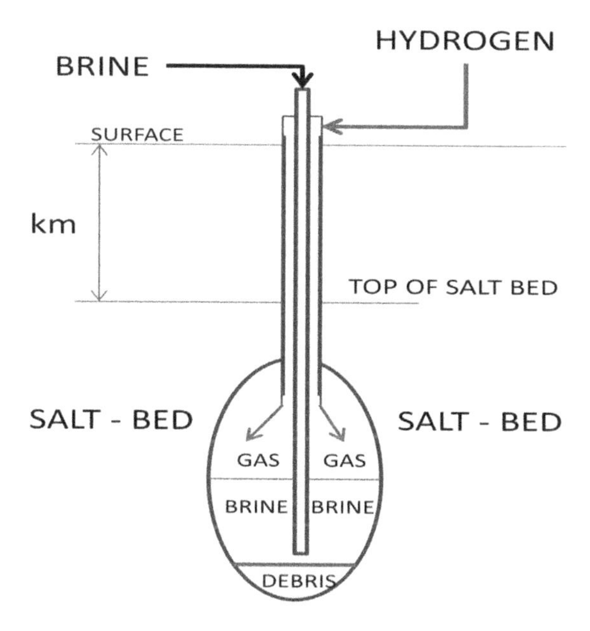

Fig. 5.5: Operation of salt cavern.

of the cavern and brine removed or added to the cavern base. For successful operation the cavern requires an adequate water supply and a method of brine disposal.

Caverns are in widespread use in the chemicals industry used for storing gaseous products such as ethylene, LPG. There are currently four caverns operating for hydrogen storage as shown in Table 5.2.

Storage as Hydrides

Over the past several decades there has been a good deal of interest in storing hydrogen as a hydride. Hydrides are of two types: (i) complex hydrides that are essentially compounds of hydrogen with other elements, and (ii) intermetallic hydrides that exploit the tendency of some metals and alloys to absorb hydrogen with the hydrogen atoms diffusing into the metal/alloy crystal.

Metal hydrides have the potential to offer very high hydrogen storage densities that can only be achieved by very high-pressure storage. They have been proposed for stationary and vehicle storage.

Table 5.2: Operating salt domes for storing hydrogen.

Country	US	US	US	UK
Location	Clemens Dome	Moss Bluff	Spindletop	Teeside
Owner	Conoco-Phillips	Praxair	Air Liquide	Sabic
Commencing	1983	2007	2014	1972
Volume (10^3)	580	566	>580	3 units at 70
Pressure (bar)	70–135	55–152	unknown	45
Energy (GWh)	92	120	120	25
Hydrogen (approx. t)	2772	3615	3615	753

J.B. von Colbe[7] *et al.* gives a recent review of the subject. Table 5.3 lists some of the systems under development for stationary storage. The table lists the hydride, the theoretical hydrogen capacity, the enthalpy of desorption and absorption, the entropy of desorption and absorption, the hydride crystal density, the theoretical hydrogen storage capacity in terms of kg H_2/m^3, the estimated cost of storage and the minimum hydrogen desorption temperature at 0.1 MPa pressure. The table shows that some of the systems have a very high storage capacity but, when detailed, at a high cost.

In complex hydride systems the hydrogen is obtained essentially by decomposing the hydride at a specific temperature. Very high theoretical storage capacities can be obtained e.g. 70.1 kg/m^3 for sodium aluminium hydride, $NaAlH_4$.

The problem is the regeneration of the starting hydride. The material left in the storage vessel cannot be recycled, rather it is taken off-site for regeneration. This is often a complex and energy intensive process. These systems would work rather like a zinc–manganese dioxide battery for delivering power.

Inter-metallic hydrides offer the potential for a reversible process. The hydrogen is absorbed at a low temperature and desorbed at a higher

[7] von Colbe JB, Ares J-R, Barale J, *et al.* (March 2019) Application of hydrides in hydrogen storage and compression: Achievements, outlook and perspectives. *Int J Hydrogen Energy* **22:** 7780.

Table 5.3: Properties of some metal hydrides.

Hydride materials	**Theory H_2 capacity (wt %)**	**ΔHdes/ ΔHabs (kJ/mol)**	**ΔSdes/ ΔSabs (J/mol.K)**	**Hydride crystal density (g/cm^3)**	**Theory (kg-H_2/m^3)**	**Cost of H_2 storage (US$/kg)**	**1 bar H_2 des temp. (°C)**
Complex Hydrides							
$NaAlH_4 \leftrightarrow \frac{1}{3}Na_3AlH_6 + \frac{2}{3} Al + H_2$	3.73	38.4/–35.2	126.3/–118.1	1.251	46.7	107.2	31
$Na_3AlH_6 \leftrightarrow 3NaH + Al + 3/2H_2$	2.96	47.6/–46.1	126.1/–123.8	1.455	43.1	120.6	104
$NaAlH_4 \leftrightarrow Na_3AlH_6 \leftrightarrow NaH + Al + 3/2H_2$	5.6	n.a.	n.a.	1.251	70.1	71.4	n.a.
$Na_2LiAlH_6 \leftrightarrow 2NaH + LiH + Al + 3/2H_2$	3.52	54.95/n.a.	135/n.a	1.405	49.4	n.a.	134
$K_2LiAlH_6 \leftrightarrow 2KH + LiH + Al + 3/2H_2$	2.56	n.a.	n.a.	1.571	40.2	n.a.	n.a.
$KAlH_4 + LiCl \leftrightarrow KCl + LiH + Al + 3/2H_2$	2.69	37.6/–37.6	97.9/–97.9	1.457	39.2	n.a.	111
$Mg(NH_2)_2 + 2LiH \leftrightarrow Li_2Mg(NH)_2 + 2H_2$	5.58	38.9/n.a	112.0/n.a	1.182	65.9	n.a.	75
Inter-metallic Hydrides							
AB_2 type — Hydralloy C5: Ti0.909Zr0.05 2Mn1.498V0.439Fe0.086Al0.016H3	1.88	28.3/–22.5	111.9/–97.2	5	93.8	n.a.	–20.7
Ti1.2Mn1.8H3	1.9	28.7/n.a	114/n.a	5.23	99.2	321.5	–21.4
TiFeH2.0 ↔ TiFeH1.04g	0.91	n.a.	n.a.	5.47	50	n.a.	n.a.
TiFeH1.04 ↔ TiFeH0.1	0.9	28.1/n.a	106.3/n.a	5.88	53.2	n.a.	

temperature. Upon cooling the system can be used again for another cycle. One advantage of this system is the desorption at the higher temperature can be performed at higher pressure. The system can thus operate as a hydrogen compressor.[8]

Clearly hydrides have advantages for some niche operations requiring hydrogen but the high cost of storage is a barrier to their widespread use for large-scale applications. After much development work for vehicle storage, the introduction of high-pressure composite storage cylinders has largely replaced work on hydrides for this purpose. However, there are many smaller uses of fuel cells for which hydrides act as a useful source of hydrogen which have been reviewed by Lototskyy *et al.*[9]

There has recently been an interest in using magnesium hydride (MgH_2) as a means of storing and producing hydrogen on a large scale. The method involves the use of nanoparticles of magnesium, doped with transition metals, which can take up hydrogen at high pressure and release it when heated. The high temperature required (typically 280°C) to release the hydrogen is a hurdle to the uptake of the method.[10]

Hydrogen Compression

In comparison to other transport fuels, hydrogen has a low volumetric energy. This is illustrated for various fuel in Figure 5.6.

As is immediately apparent, the conventional fuels have a very high volumetric energy density that delivers to them their competitive advantage over alternatives; these include lithium ion batteries that have energy densities in the range of 0.9–2.63 MJ/L.[11] For practical use the hydrogen has to be used at the highest pressure possible, preferably at about 800 bar when it has a volumetric energy density of about 10 MJ/L. Taking into consideration the improved energy efficiency of fuel cells over

---

[8] *ibid.*

[9] Lototskyy MV, Tolj I, Pickering L, Sita C, Barbir F, Yartys V. (2017). The use of metal hydrides in fuel cell applications. *Prog Nat Sci Mater Int* **27:** 3–20.

[10] Schlapbach L, Zuttel A. (2002) Hydrogen-storage materials for mobile applications. *Nature* **414:** 353; Song Z, *et al.* (2 July 2020) Enhancing Hydrogen Storage Properties of MgH_2 by Transition Metals and Carbon Materials: A Brief Review. *Front Chem.*

[11] From https://en.wikipedia.org/wiki/Energy_density

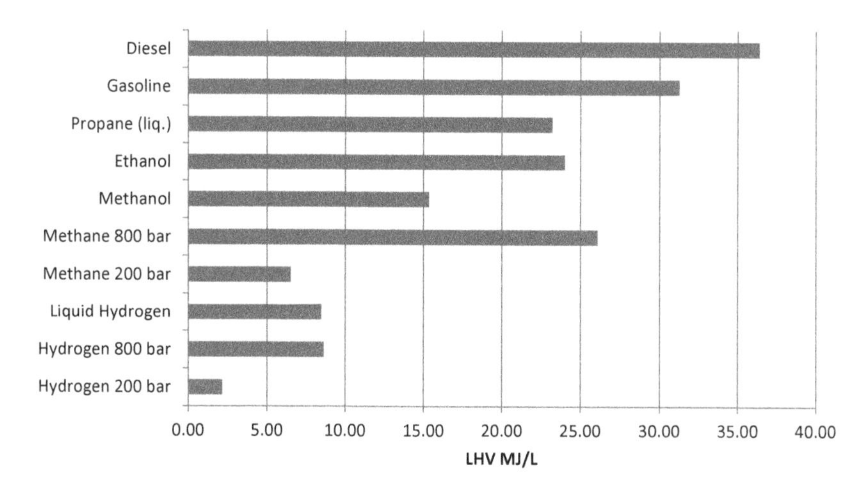

Fig. 5.6: Volumetric energy density for various transport fuels of interest.

conventional internal combustion engines this is equivalent to about 15 MJ/L that is similar to liquid methanol.

There is a practical limit to increasing pressure because although the increase in density is fairly proportional to pressure up to about 30 MPa (300 bar), further increasing pressure has less effect[12]; 70–80 MPa (700–800 bar) seems to be a working optimum.

The adiabatic compression of hydrogen requires considerably more specific energy than the adiabatic compression of methane. The preference for hydrogen compression is isothermal, which requires considerably less energy. This is illustrated in Figure 5.7 for the ratio of the compression energy required to the higher heating value (HHV) of hydrogen[13] under adiabatic and isothermal conditions.

In practice hydrogen compressors are of the multi-stage type with intermediate cooling with the pertinent curve lying between the curve for adiabatic and isothermal. The actual position of the curve will depend on the design of the individual compressor system. The result is that

---

[12] Makrides SS. (1976) *Encyclopedie de Gaz, Air Liquide, Division Scientifique*. Elsevier, Amsterdam, Netherlands.

[13] Adapted from Makrides quoting Bossel U, Eliasson B, Energy and the hydrogen economy. US DoE Office of Energy Efficiency and Renewable Energy (EERE). http://www.afdc.energy.gov/pdfs/hyd_economy_bossel_eliasson.pdf

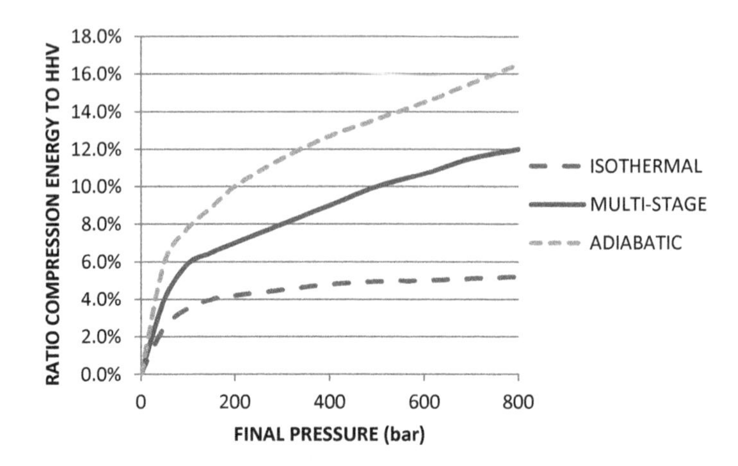

Fig. 5.7: Ratio of compression energy to HHV of hydrogen at various final pressures.

Fig. 5.8: Compression power required for a central station delivering 373 kt/y hydrogen.

compression of hydrogen from 0.1 MPa (1 bar) to 80 MPa (800 bar) will require the equivalent of 10% of the energy content of the hydrogen being compressed.

The cost of hydrogen compression is further illustrated in Figure 5.8 that estimates the compression power required against delivery pressure for a central facility producing 373 kt/y hydrogen assuming hydrogen is available at 3 MPa (30 bar), e.g. exit of a steam reformer or gasifier. For delivery at 80 MPa (800 bar) for vehicle filling requires over 100 MW.

For small-scale (such as domestic) operations the compressors of choice are probably based on a diaphragm design or possibly an electrical proton exchange membrane compressor. For larger scale operations the multi-stage reciprocating piston compressors would be preferred. One of the problems with many of these types of compressors is the tendency to contaminate the hydrogen with compressor oil. This requires removal with appropriate filters or oil traps prior to use.

Hydrogen Transport by Pipelines

Pipelines are well known for the transport of hydrocarbon fluids, both oil and gas. They are widely used for the transport of other chemicals including hydrogen. Within large refinery and chemical operations, pipelines ferry hydrogen from one unit operation to another, often over several kilometres. What we are mainly concerned about here is the mass transport of hydrogen over hundreds of kilometres to facilitate the distribution of hydrogen from a remote production site. Such hydrogen pipelines are well known with over 2000 km of pipelines in the United States and over 250 km of pipelines in Europe.

Table 5.4 shows that most of the world's hydrogen pipelines are in the United States and Europe and are in the main operated by the major merchant gas suppliers. The earliest reported hydrogen pipeline is the Rhine–Ruhr pipeline built 1938 and said to be still in operation.

Compared to natural gas pipelines, there are several issues with hydrogen pipelines[14]:

- Hydrogen promotes embrittlement of the pipeline
- Hydrogen induces corrosion
- Hydrogen pipeline ruptures **always** catch fire
- A flare is required at pipeline vents and relief valves

[14] AirLiquide, US DoE Hydrogen Pipeline Working Group Meeting, 31 August 2005.

Table 5.4: World hydrogen pipelines.[15]

Company	km
Air Liquide	1936
Air Products	1140
Linde	244
Praxair	739
Others	483
World Total	**4542**
U.S.	2608
Europe	1598
Rest of world	337
World total	**4543**

- Small leaks are hard to detect
- Odourisation is difficult
- Construction costs for hydrogen pipelines are higher

These issues are compounded because hydrogen can diffuse within the crystal lattice of most metals and hence diffuse out of the pipeline leading to hydrogen loss.

In order to avoid embrittlement high-strength steels should be avoided. For high-pressure pipelines the preferred construction material is micro-alloyed steels (API 5L X52 grade) that have been used at pressures of 70 MPa.[16]

In the United States some steel, oil pipelines have been converted to carry hydrogen. These operate typically at a maximum pressure of 5 MPa (50 bar). The use of reinforced thermoplastic pipe (Soluforce™[17]) has been demonstrated at Groningen Seaports.

[15] Anon. www.h2tools.org. 'Hydrogen Pipelines September 2016.xlsx': this spreadsheet details a large number of the world's hydrogen pipelines.

[16] The detail of the construction of hydrogen pipelines including the materials of construction are given in *European Industrial Gases Association*, Hydrogen Transportation Pipelines, IGC Doc 121/04/E.

[17] Soluforce is the brand name of Pipeline Netherlands B.V a subsidiary of Wiernerberger AG.

Being a gas hydrogen is a compressible fluid and so long-distance pipelines would generally require recompression at appropriate intervals. This is well known for natural gas pipelines where part of the flow is intercepted to run the recompression station. This may not be possible for hydrogen and recompression may require a different fuel (e.g. diesel) at the recompression station.

For a central facility (373 kt/y) passing all of the hydrogen down a single line to a distant distribution centre would in most cases require recompression stations — the number depending on the pipeline diameter. Scoping estimates indicate that a large 36-inch pipeline may be able to pass the required volume 1600 km without recompression but, as noted later, pipeline cost is a function of pipe diameter and so would increase the capital cost of the pipeline.

Use of Existing Natural Gas Pipelines

In many parts of the world there are large networks of natural gas pipelines. There are several proposals to use these networks for the transport of hydrogen. In these schemes hydrogen is blended into the natural gas stream at one or more points and separated from the natural gas at another point in the system. Clearly the method has merit in saving the high capital cost of building a dedicated pipeline.

However, there are some issues that result on blending hydrogen into natural gas. There are potential high losses from seals and fittings that are designed for natural gas rather than hydrogen. The hydrogen may cause steel embrittlement in the line because the steel would not be of the appropriate quality. However, as noted above, some steel pipelines have been converted for hydrogen carriage, embrittlement may be less of an issue if the partial pressure of the hydrogen in the line is low.

The extraction of the hydrogen at the user end would require additional process plant. Some natural gas lines are used to transport LPG with the LPG extracted in a so-called straddle plant. This duty is performed by cooling the gas stream to condense out the LPG. This is not possible for hydrogen but it would be feasible to use pressure swing adsorption (PSA), cryogenic or membrane separators which are well

Table 5.5: Pertinent properties of hydrogen and methane.

Property	Unit	Hydrogen	Methane
Higher heating value (HHV)	MJ/m3	12.1	37.7
Lower heating value (LHV)	MJ/m3	10.2	34.0
Specific gravity (air = 1)		0.0696	0.5539
Higher Wobbe Index		45.9	50.7
Lower Wobbe Index		38.7	45.7
Methane Number		0	100

known in the process industries for recovering hydrogen in refinery waste gas streams.[18]

However, probably the biggest issue that defines the upper limit for blending hydrogen into natural gas is the impact on the gas specification. This is of particular concern for common carrier pipeline networks that may interconnect many gas producers and users. Such systems have tight gas specifications and regulations, the detail depending on the location and the jurisdiction. These specifications limit the maximum hydrogen content to about 20% and for many systems it will be lower than this.

A gas line specification covers a range of parameters but we are mainly concerned with the heating value and the Wobbe Index. Also, to be considered is the methane number of the gas. Methane number is usually not specified in gas pipeline systems but is of considerable interest to customers who use gas-driven engines for uses such as power generation. Table 5.5 compares some pertinent values of hydrogen and methane for pipeline transport.

Heating value

We are primarily concerned with the heating value expressed in volumetric terms. The problem for hydrogen is that the heating value when expressed in volumetric terms is less than one third that of methane.

Natural gas is universally sold on an HHV basis and most jurisdictions use the HHV for natural gas pipeline specification. Sometimes there

[18] Faraji S, Sotudeh-Gharebagh S, Mostoufi N. (2005) Hydrogen Recovery from Refinery Off-gases. *J Appl Sci* **5:** 459.

are multiple specifications for a given line, e.g. when it has the duty to carry LPG.

Methane is the main component of natural gas and has a HHV of 37.7 MJ/m^3 when measured at 15°C compared to hydrogen at only 12.1 MJ/m^3. Hydrocarbon components higher than methane have higher calorific values (volumetric), for instance ethane 66.7 MJ/m^3 and propane 93.9 MJ/m^3. Sometimes these components of natural gas are left in the gas to lift the overall calorific value to over 40 MJ/m^3. This is common in the Asia-Pacific region. In other regions these higher components are stripped out of the gas and the natural gas has a correspondingly lower calorific value, nearer to that of methane. Nevertheless, blending large volumes of hydrogen into a natural gas network will lower the calorific value.

Wobbe index

In some jurisdictions the specification is determined by the Wobbe Index (WI). The use of the WI is to determine if two gases of differing composition can be used for the same combustion duty.

WI is defined as:

$$\mathbf{WI = Heating\ Value/\sqrt{(specific\ gravity)}}$$

Two WIs can be defined depending whether or not the heating value is measured on a HHV or a LHV basis. There is a marked difference between methane and hydrogen in specific gravity as well as the heating value: hydrogen s.g. 0.0696, methane s.g. 0.5539. The higher WI for hydrogen is 45.9 compared to methane at 50.7.

If a gas is being used for a particular duty, then substitution of the gas with another of higher WI can lead to overheating of the burner, carbon monoxide formation, soot build-up and, potentially, an explosion. If the substitute gas has a lower WI, then flash-back at the burner can occur, the stability of the flame can become a problem and there can be difficulty in igniting the system.

Blending hydrogen into a natural gas network will disturb the WI of the gas and may cause serious problems for downstream users. The use of a natural gas pipeline to carry hydrogen may not be a problem if there are few users and the hydrogen is removed prior to the widespread distribution through a pipeline network servicing many users. In other words, if the pipeline is for example a long-distance pipeline with few, if any, users of the gas along the way.

Methane number

The methane number of a natural gas applies specifically to the use of natural gas in gas engines. The number is widely used by gas engine manufacturers. At the time of writing, there is no agreed method of measurement and the determination varies with different engine manufacturers. Often the methane number is computed from the H/C molar ratio of the gas. Methane number operates like octane (which is measured in a defined gasoline engine) and is measured on each manufacturer's particular engines.

The methane number of a gas is equal to the performance of the engine operating with a mixture of methane and hydrogen. Methane is given the methane number of 100 and hydrogen the methane number of zero. A gas comprising 80% methane and 20% hydrogen would this have a methane number of 80.

Gas engine suppliers specify the minimum methane number of the gas to be used for optimum performance of their engine. Clearly adding hydrogen to a natural gas pipeline system could lower the methane number of the gas reducing the performance of associated gas engines. The problem becomes apparent for long-distance pipelines through remote areas when often a spur line delivers gas to a remote operation, facility or township that rely on gas engines to generate electricity. Lowering the methane number to below the engine manufacturers specification would require the de-rating of the engine — i.e. produce less power for a given volume of gas.

Pipeline Costs

The cost of construction of a pipeline is a function the pipeline materials, labour to install the line, miscellaneous materials and the cost of the access to the pipeline easement, the Right-of-Way.[19] A typical breakdown of costs for a land-based pipeline is illustrated in Figure 5.9.

[19] The breakdown of US pipeline costs is reported annually by the *Oil & Gas Journal*.

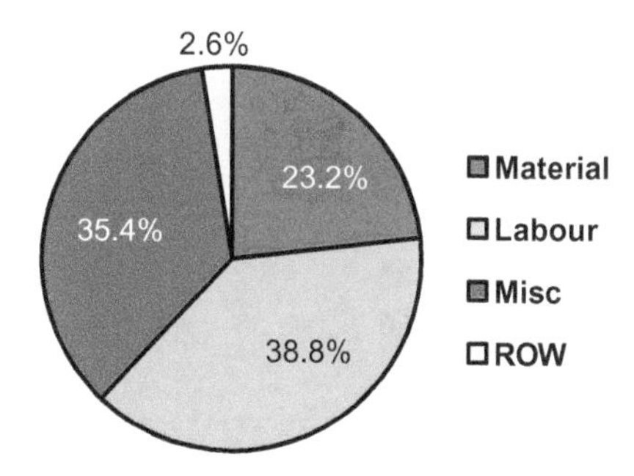

Fig. 5.9: Typical breakdown of pipeline costs.

The cost of the pipe itself with the right-of-way (ROW) costs are only about 25% of the final installed cost. The ROW varies considerably with location. For pipelines running through open country, it can be quite low but for pipelines running through urban areas this cost can be considerable. For pipelines running through several jurisdictions (country borders) additional ROW charges could be incurred, e.g. metering stations so that royalties can be charged on the volume flowing through the pertinent jurisdiction. The material cost of the pipeline is a function of the pipe diameter (usually measured in inches) and length (measured in miles or kilometres). For hydrogen pipelines using more expensive steels the materials cost would be higher.

The miscellaneous charges average about 35% of the final cost. The miscellaneous cost covers valves and fittings, pig launching stations for inspection and cleaning, pipeline supports and the like. For long pipeline systems the miscellaneous charges also cover compressor stations. For a fully compressed line (maximum flow) there could be a compressor station about every 60 km or so. A major additional cost incurred in the miscellaneous section is river crossings. These usually have to be individually designed, cost obviously depends on the size and flow of the river. They rarely cost less than about $2 million each, for some long-distance pipelines there may be many rivers to cross.

The labour cost at nearly 40% of the total, encompasses all of the costs of laying the pipeline and installing the miscellaneous fittings, compressor stations, river crossings etc.

Figure 5.9 is a typical average for US land-based pipelines. Most pipeline construction companies have generalised formulae for estimating

Table 5.6: The estimated cost of some long distance European natural gas pipelines.

Pipeline	Setting	km	Diameter inches	Flow Bm^3/y	Capex B€	M€/ km	M$/ km[a]	$/in/km
Nord Steam 1	Offshore	1224	48	55	8.8	3.59	4.03	83,878.0
Nord Stream 2	Offshore	1230	48	55	10.0	4.07	4.55	94,850.9
Gryazovets-Vyborg	Onshore	917	56	55	4.5	4.91	5.50	98,146.1
Gryazovets-Ust Luga	Onshore	920	56	55	3.2	3.48	3.90	69,565.2
NEL	Onshore	440	56	20	1.0	2.27	2.55	45,454.5
OPAL	Onshore	470	55	35	1.0	2.13	2.38	43,326.9
EUGAL	Onshore	480	56	55	4.0	8.33	9.33	1,66,666.7
Gazelle	Onshore	166	56	33	0.4	2.41	2.70	48,192.8
GIPL	Onshore	562	28	2.3	0.6	1.07	1.20	42,704.6
GIPS	Onshore	164	39	5.7	0.3	1.83	2.05	52,532.8
Baltic Pipe	Offshore	275	36	10	0.4	1.45	1.63	45,252.5

[a] Euro = 1.12US$ (2019 average)

pipeline costs usually based on currency/inch (diameter)/km (length). For example, onshore pipelines in Europe are reported[20] to cost Euro 35,000 to 45,000/in/km. Offshore pipelines can cost double that of onshore pipelines. The average US pipeline cost is M$3.2/km at 2018 cost basis.

The estimated cost of some long-distance natural gas pipelines in Europe are given in Table 5.6.[21] Onshore lines are compressed, offshore lines are uncompressed.

Also of note is the reported typical pipeline tariff of $15/1000 m^3/1000 km. In many jurisdictions pipeline tariffs are highly regulated and many use the heating value of the gas (i.e. natural gas) transported as their basis rather than the volume of gas. It is debateable how hydrogen carried will be charged by pipeline owners.

For some jurisdictions, such as Australia, gas pipeline cost is considerably lower than the reported US averages and the European lines.

---

[20] Przybylo P. (4 May 2020). Nord Stream 2 project advanced despite weak economics. *Oil Gas J* 46.

[21] *ibid.*

Achievement of the lower cost base is facilitated by pipelines passing long distances through open country with few river crossings or similar obstacles.

Cost of Hydrogen Storage and Transport

Amos WA[22] produced an extensive analysis for the storage and transport of hydrogen including in the large volumes required for a central facility in the United States. Although the data is somewhat old (1998) its comprehensive nature and analysis is used to build current estimates for storage and transport in this work. In 1998 the use of metal hydrides was an active proposal for the storage, transport and vehicular use of hydrogen. Since that time the use of metal hydrides has fallen out of favour for large-scale vehicular transport and is not discussed in this section.

Cost of Hydrogen Storage

Three cases are considered for storage — compressed gas, liquid hydrogen and caverns. The capital costs of the estimates include the cost of hydrogen compressors for the required duty, which is typically compression from 0.1 MPa to 20 MPa. The amount of storage required is for 30 days production.

Compressed gas

The data for this case is presented in Table 5.7. The top half of the table reports the original data that has been converted as necessary to SI units. The case is for the cost of storage of hydrogen as compressed gas in an appropriate pressure vessel or vessels at 20 MPa with a capacity to hold 30 days of production.

The bottom half of the table transfers the data to the central hydrogen production facility used as the basis in this work and inflates cost to mid-2018 values. The unit cost of hydrogen storage is then estimated by the

[22] Amos WA. (November 1998) *Cost of Storing and Transporting Hydrogen*. NREL/TP-570-25106.

Table 5.7: Cost of hydrogen storage as compressed gas in a tank (after Amos).

Rate of hydrogen production	kg/h	45,359
	kt/y	370.1
	kt/d	1.08
Storage	days	30
Method		TANK
Pressure	MPa	20
CAPEX (1998)	M$	2270
Compression power	kW	100,000
Cooling water	L/h	2,271,240
	m^3/h	2271.24
Central facility	size ratio	1.01
Capital cost (2018)	M$	4435.22
ROC (20 y life, 1 y con; 10% DCF)	% CAPEX	12.92
ROC	M$/y	573.06
Power required	MW	101.03
Power price	$/MWh	50.00
	M$/y	41.22
Cooling Water Cost	$/$m^3$	0.02
Cooling Water Cost	M$/y	0.37
TOTAL hydrogen storage cost	M$/y	614.65
	$/t	1643.71
	$/GJ (LHV)	13.76

methodology used in this work. It is assumed that the storage tank can be constructed in 1 year, has a 20-year life and requires a discounted cash flow return of 10%/year.

The analysis indicates that the cost of hydrogen storage is $1644/t or $13.76/GJ, measured on a lower heating value basis. The analysis indicates that the greater portion of this cost is the result of capital charges, which is very high for this method of storage.

Liquid hydrogen

The estimate for the storage of liquid hydrogen follows in the same manner as that for the compressed hydrogen gas. The cost is for the storage of hydrogen as liquid hydrogen requires a large increase in the compression power required. For this case it is assumed a 2-year construction period is necessary.

The cost estimate is developed in Table 5.8. The cost estimate is for the hydrogen production from a central facility requiring 30 days storage

Table 5.8: Cost of hydrogen storage as liquid (after Amos).

Rate of hydrogen production	kg/h	45,359
Actual (with boil off)	kg/h	46,700
	kt/y	381.07
	kt/d	1.1208
Storage	days	30
Method		Cryogenic tank
Boil off rate	%/day	0.1
Capital cost (1998)	M$	663.73
Compression power	kW	463,300
Cooling water	L/h	29,229,566
	m^3/h	29,229.6
Central facility	size ratio	0.98
Capital Cost (2018)	M$	1259.58
ROC (20 y life, 2 y const., 10% DCF)	%	13.57%
ROC	M$/y	170.88
Power required	MW	454.63
Power price	$/MWh	50.00
	M$/y	185.49
Cooling Water	$/$m^3$	0.02
Cooling water cost	M$/y	4.68
TOTAL hydrogen storage cost	M$/y	361.05
	$/t	965.53
	$/GJ (LHV)	8.08

as liquid hydrogen. For liquid hydrogen there is a finite boil-off (loss) rate (0.1% of stored liquid hydrogen per day) that is allowed for.

The estimate is developed in the same manner as above, which results in the storage cost being estimated at $965.53/t or $8.08/GJ (LHV basis). Note this is considerably lower than the estimate for storage as compressed gas, because the capital cost of the storage vessel for the 30-day storage required is lower.

Note the production of liquid hydrogen is discussed more fully in Chapter 6, which may indicate that the costs estimated here is low.

Underground caverns

The estimated cost for hydrogen storage in caverns is developed in Table 5.9. The underground storage option delivers lower capital cost and

Table 5.9: Cost of hydrogen storage in underground caverns (after Amos).

Hydrogen production rate	kg/h	45,359
Actual	kg/h	45,359
	kt/y	370.13
	kt/d	1.089
Storage	days	30
Method		Cavern
Losses	%/day	0
Capital cost (1998)	M$	340.53
Compressor power	kW	100,000
Cooling water	L/h	2,271,240
	m^3/h	2271.2
Central facility	Size ratio	1.01
Capital cost (2018)	M$	665.34
ROC (20 y life, 2 y const., 10% DCF))	%	13.57%
ROC (20 y, 2 y)	M$/y	90.26
Power required	MW	101.03

Table 5.9: (*Continued*)

Power price	\$/MWh	50.00
	M\$/y	41.22
CW	\$/m^3	0.02
CW cost	M\$/y	0.37
TOTAL	M\$/y	131.86
	\$/t	352.62
	\$/GJ (LHV)	2.94

similar compression power to that for above-ground storage in pressure vessel. The option estimate is based on a cavern constructed on a 2-year period, with a lifetime of 20 years delivering a 10% discounted cash flow on the capital expenditure. This option is assumed to have no losses. This results in a low cost for hydrogen storage of 352.6/t or \$2.94/GJ (LHV basis) principally as savings to the capital costs.

Summary

In summary, the cost of hydrogen storage is:

- As compressed gas in pressure vessels: \$1644/t (\$13.76/GJ, LHV)
- As liquid hydrogen: \$965.5/t (8.08/GJ, LHV)
- As compressed gas in underground caverns: \$352.6/t (\$2.95/GJ, LHV)

For large volumes of gas for long-term storage caverns are by far the cheapest method. The liquid hydrogen option would could be viable for large quantities of gas in areas of low power cost or areas requiring liquid hydrogen as a product. Note this may be a low-cost estimate. Storage as compressed gas may be useful for smaller quantities of gas requiring shorter storage times.

Amos[23] also develops cases for the use as hydrides for storing hydrogen. This is a high-cost option but is suitable for small applications.

---

[23] *ibid.*

The Cost of Hydrogen Transport

Amos develops the cases for the transport of hydrogen by various means. These have been updated as above and the data presented here. Again, the focus is on large-scale hydrogen transport scaled for a central production facility (373 kt hydrogen per year).

Cost of transporting compressed hydrogen by truck

This method envisages hydrogen transport in pressure vessels (cylinders) on trucks and trailers. This method is in widespread use. The cost of compression is not considered and is assumed available at sufficient pressure to fill the cylinders. The cases are developed for one-way distances ranging from 1609 km (1000 miles) down to 80 km. The data is given in Table 5.10.

Table 5.10: Cost of compressed hydrogen transport in cylinders (after Amos).

GAS by truck	**Distance**	**1609 km**	**805 km**	**322 km**	**80 km**
Rate of hydrogen production	kg/h	45,359	45,359	45,359	45,359
	kt/y	370.12	370.12	370.12	370.12
	kt/d	1.088	1.088	1.088	1.088
Method		Truck	Truck	Truck	Truck
Distance	km	1,609	805	322	80
Capacity	kg/truck	181	181	181	181
Trips	per year	2,100,000	2,100,000	2,100,000	2,100,000
	km/y	6.76E+9	3.38E+9	1.35E+09	3.38E+08
Time/trip	h	40	20	8	2
Total drive time	h/y	84,000,000	42,000,000	16,800,000	4,200,000
Trucks	No	10,500	5,500	2,500	1,000
Drivers	No	21,000	11,000	5,000	2,000
Fuel	ML/y	2,649.79	1,324.89	529.96	132.49
Capital cost (1998)	M$	2,625	1,375	625	250
Central facility	Size ratio	1.01	1.01	1.01	1.01

Table 5.10: (*Continued*)

GAS by truck	Distance	1609 km	805 km	322 km	80 km
Capital cost (2018)	M$	5,128.84	2,686.53	1,221.15	488.46
ROC (10 y life, 1 y con, 10%)	% Capex	17.90%	17.90%	17.90%	17.90%
Return on capital	M$/y	918.16	480.94	218.61	87.44
Fuel	ML/y	2,649.79	1,324.89	529.96	132.49
Fuel price	$/L	1.00	1.00	1.00	1.00
	M$/y	2,649.79	1,324.89	529.96	132.49
Driver	$/y	120,000	120,000	120,000	120,000
	M$/y	2,520.00	1,320.00	600.00	240.00
TOTAL	M$/y	6,087.95	3,125.84	1,348.57	459.93
	$/t	16,280.55	8,359.19	3,606.37	1,229.96
	$/GJ (LHV)	136.27	69.97	30.19	10.29

The data is presented in a like manner to that for hydrogen storage above. The top portion of the table is based on the original data and the bottom half is updated to 2018 costs in order to estimate the cost of hydrogen transport.

The cost of transport is evaluated over several distances ranging from 1609 km (1000 miles) to 80 km. Each truck trailer system is assumed to carry 181 kg of hydrogen. For carrying gas from a central facility 370 kt/y would require 2.1 million trips per year. For a 1609-km one-way journey, this amounts to a total of 6.78 billion kilometres travelled. This will require 10,500 truck/trailers and 21,000 drivers. By contrast moving the same volume of gas over a one-way distance of 80 km would only require 1000 trucks and 2000 drivers.

The assumptions are that each truck costs $459,000, has a 10-year life, takes 1 year to construct and requires a 10% discounted cash flow return on the capital expended, with fuel prices at $1/L and a driver cost (including overheads) of $120,000/year. This results in the cost of hydrogen transport over 1609 km being $16,280/t and over 80 km this falls to $1229/t or $10.29/GJ (LHV). Clearly the use of truck/trailer transport would be limited to relatively short distances.

Cost of transporting liquid hydrogen by truck

In a similar manner the cost of transporting liquid hydrogen by truck is developed with the estimates given in Table 5.11. The cost of liquefaction is not considered.

Table 5.11: Estimated cost of delivering liquid hydrogen by truck (after Amos).

Liquid hydrogen by truck		**1609 km**	**805 km**	**322 km**	**80 km**
Rate of hydrogen production	kg/h	45,359	45,359	45,359	45,359
	kt/y	370.13	370.13	370.13	370.13
	kt/d	1.088	1.088	1.088	1.088
Method		TRUCK	TRUCK	TRUCK	TRUCK
Distance	km	1609	805	322	80
Capacity	kg/truck	4082	4082	4082	4082
Trips	per year	93,333	93,333	93,333	93,333
	km/y	1.87E + 08	9.33E + 07	3.73E + 07	9.33E + 06
Time/trip	h	42	22	10	4
Total drive	h/y	8.40E + 07	4.20E + 07	1.68E + 07	4.20E + 07
Hydrogen boil off rate	%/day	0.30%	0.30%	0.30%	0.30%
Delivered	kt/y	368.19	369.11	369.67	369.94
Trucks	No	467	245	112	45
Drivers	No	934	489	223	89
Fuel	ML/y	117.77	58.88	23.55	5.89
Capital cost (1998)	M$	233.5	122.5	56	22.5
Central facility	scale	1.02	1.02	1.02	1.02
CAPEX (2018)	M$	458.63	240.61	109.99	44.19
ROC (10 y life, 1 y con, 10 % DCF)	%	17.90%	17.90%	17.90%	17.90%
Return on capital	M$/y	82.10	43.07	19.69	7.91
Fuel	ML/y	117.77	58.88	23.55	5.89
Fuel price	$/L	1.00	1.00	1.00	1.00
	M$/y	117.77	58.88	23.55	5.89

Table 5.11: *(Continued)*

Liquid hydrogen by truck		1609 km	805 km	322 km	80 km
Driver	$/y	1,20,000	1,20,000	1,20,000	1,20,000
	M$/y	112.08	58.68	26.76	10.68
TOTAL	M$/y	311.95	160.64	70.00	24.48
	$/t	834.23	429.58	187.20	65.46
	$/GJ (LHV)	6.98	3.60	1.57	0.55

For the case of transporting liquid hydrogen, the truck requires a cryogenic storage vessel (Dewar) that has a high capital cost due to the quantity and quality of the insulation required. The capital cost of the vehicle is double that for the compressed gas case. However, each vessel carries 4 t of hydrogen. There is some boil-off that is allowed for.

The assumptions are that each truck costs $980,000, has a 10-year life, takes 1 year to construct and requires a 10% discounted cash flow return on the capital expended, with fuel prices at $1/L and a driver cost (including overheads) of $120,000/year.

The higher cost of the trucking fleet is offset by the carriage of more hydrogen per trip. The estimated cost of liquid hydrogen transport for a 1609-km one-way distance is $834/t or $6.98/GJ (LHV) and for a 80-km distance the estimated cost of liquid hydrogen transport by truck is $65.5/t or $0.55/GJ (LHV). Clearly this is considerably less than the comparative cases using compressed gas cylinders.

Cost of transporting liquid hydrogen by barge

Amos also considers the case of mounting the liquid hydrogen Dewar vessels on a ship for transport. The case presented here is for a 1609-km one-way journey, which is rather short for trans-oceanic shipments, but could be relevant to operation in the US Gulf or the North Sea and the associated large river systems. The case developed is for transport of the liquid hydrogen in a large Dewar-type container that is placed aboard a ship or barge. Transport in large volumes of liquid hydrogen in specially constructed ships over trans-oceanic distance is considered in Chapter 6. The case considered here is detailed in Table 5.12.

This case considers placing the Dewar vessels on board a ship or barge and paying the appropriate carriage costs, which in the original

Table 5.12: Transport of liquid hydrogen by ship (after Amos).

Liquid hydrogen by ship		**1609 km**
Hydrogen production rate	kg/h	45,359
	kt/y	370.13
	kt/d	1.08
Method		SHIP
Distance	km	1,609
Capacity	kg/tank	4,082
Tanks	per year	93,333
	km/y	187E + 6
Tanks required		3,200
Time/trip	h	240
Boil off rate	%/day	0.30%
Delivered	kt/y	359.03
Tanks	No	3,200
Cost/tank	$	350,000
Capital cost	M$	1,120.00
Shipping cost	$/tank	3,000
Central facility		
Scaling		1.042
CAPEX (2018)	M$	2,255.98
ROC (15 y life, 2 y con, 10%)	%	15.66%
Return on capital	M$/y	353.31
Shipping cost	$/tank	5,801.80
	M$/y	541.50
TOTAL	M$/y	894.81
	$/t	2,492.34
	$/GJ (LHV)	20.86

paper were estimated at $3,000/tank, inflated to $5,800/tank for 2018. For the 1609-km journey, 3200 tanks are required, which results in a high capital cost for this scheme. The costs are also higher due to consideration of the turnaround time for loading and unloading the vessel and the slower rate of travel (16 km/h; typical of river speeds). The estimated hydrogen transport cost is $2,492/t or $20.86/GJ (LHV).

Hydrogen transport by pipeline

The cost for transporting hydrogen by dedicated pipeline over 1609 km are given in Table 5.13.

The pipeline modelled is 0.91 m (36 inch) in diameter with an inlet pressure of 20 MPa and an outlet pressure of 14 MPa. Hydrogen compression is required at the start of the pipeline but there are no intermediary

Table 5.13: Transport of hydrogen by pipeline (after Amos).

Rate of hydrogen production	kg/h	45,359
	kt/y	370.13
	kt/d	1.088
Method		PIPELINE
Distance	km	1609
Diameter	m	0.25
Inlet pressure	MPa	20
Compressor power	kW	73,744
Power required	kWh/y	619,450,278
Compressor cost	M$	54.96
Pipeline	M$	1,000.00
Capital cost (1998)	M$	1,054.96
Central facility		
Scaling		1.010
Capital cost (2018)	M$	2,040.22
ROC (20 y life, 2 y const., 7.5%)	%	10.94
Return on capital	M$/y	223.21
OPEX (2% Capex)	M$/y	40.80

(Continued)

Table 5.13: *(Continued)*

Power cost	\$/MWh	50.00
	M\$/y	30.97
TOTAL	M\$/y	294.98
	\$/t	788.85
	\$/GJ (LHV)	6.60

re-compression stations. Scaling of the capital cost to 2018 values gives a total capital expenditure of just over \$2 billion (see above discussion of pipeline costs). The rate of return of capital (ROC) on pipelines is often regulated, especially if access to third parties is allowed (common carrier). For this reason, the ROC discounted cash flow required is placed at 7.5% per annum rather than the more usual 10%. This results in the estimated cost for the delivery of hydrogen by pipeline of \$788/t or \$6.60/GJ.

Summary

For transport of large volumes of hydrogen produced at a remote central facility the cost of transport as compressed gas by truck is \$16,280/t (136.3/GJ LHV), for liquid hydrogen by truck is \$834/t (\$7.0/GJ), in Dewar-type vessels on a ship or barge the cost is \$2492/t (\$20.9/GJ, LHV) and by pipeline \$789/t (\$6.60/GJ, LHV).

Amos's comprehensive work on this subject also considers and evaluates many other variations on this theme such as:

- Rail transport
- Transport as metal hydrides
- Varying rates of production, down to very small scale
- Varying distances

The original work should be consulted for these scenarios.

CHAPTER 6

THE MASS SHIPPING OF HYDROGEN

One of the principal aims of this book concerning the hydrogen economy is the production and distribution of large volumes of hydrogen in advanced economies. If this comes to fruition, for a variety of reasons many economies will be unable to produce sufficient hydrogen within their borders and consequently may become major importers of hydrogen. For continental jurisdictions import by pipeline may be feasible, however, for many countries this will not be possible and it will become necessary to import hydrogen by ship. This raises the question of the manner of import either as gas (compressed gas) or liquid hydrogen — or by means of a carrier fluid. In the last-mentioned mode of transport a hydrogen-rich fluid is synthesised where the hydrogen is available and carried by ship to an importing region where the hydrogen is extracted from the carrier fluid. There are two main propositions for the carrier fluid — ammonia or a naphthene (cyclic hydrocarbon). This chapter will discuss the technologies and estimate the shipping costs as would be applied to various alternative scenarios.

Cost of Ships and Shipping

The Shipping Fleet

In the energy fuels and chemicals market there exists a large merchant marine dedicated to transport of these commodity products. The fleet is split between dedicated vessels and a large contract fleet of many different types and carrying capacity and. The contract fleet is owned by third parties who contract out individual vessels on a short- or long-term basis. The dedicated fleet of ships is owned and operated by a single owner for the carriage of specific materials, often over specific routes. The dedicated fleet is best understood by reference to the production and export of liquefied natural gas (LNG). The shipping fleet is owned and operated by the owners of a particular LNG production facility. Sometime ago, dedicated fleets were common in the oil industry with the major oil companies owning their own vessels for the supply of crude oil to their refinery operations. These dedicated fleets have been largely replaced by contract fleets although some of the oil majors still have their own dedicated shipping operations (e.g. Shell, BP).

The outline of the energy fuels shipping fleet is described in Table 6.1. The table sets out the types of products being shipped against the character of the shipping fleet.

Transported LNG is shipped in large vessels containing cryogenic tanks holding liquid methane at −165°C.[1] There are several types of tank in use, e.g. spherical or membrane. Cargo sizes range from 50,000 m^3 to over 200,000 m^3, typically 50,000–100,000 DWT. The LNG fleet has

Table 6.1: Outline of the energy fuels and chemicals shipping fleet.

Products	LNG	LPG	Chemicals	Clean fuels	Dirty fuels
Products shipped	LNG and ethane	LPG, ammonia, chemicals	Liquid chemicals	Naphtha, gasoline, gas oil (diesel)	Crude oil, fuel oil
Size (DWT)	50,000–100,000	10,000–75,000	10,000–40,000	60,000–120,000	120,000 to >250,000
Ship types	Cryogenic	Pressurised and cryogenic	Sealed tanks	Sealed tanks	Sealed tanks
Fleet	Dedicated	Contract	Contract	Contract and dedicated	Contract and dedicated
Cost variation	Fixed	Seasonal	Business cycle	Business cycle	Business cycle

[1]GIIGNL (The International Group of Liquid Natural Gas Importers), LNG Information paper No 3.

grown rapidly over the past decades as LNG has become of increasing interest as fuel for base-load power generation. LNG operations involve considerable capital investment, a significant part of this investment is often a dedicated fleet of LNG carriers or in some cases contract vessels on a long-term contract. The LNG-type vessels can also be used to carry ethane and this type of vessel could potentially be used to transport liquid hydrogen (at −253°C). Whether or not the current LNG carriers could be used to accommodate liquid hydrogen is as yet unknown but it should be feasible to place large spherical liquid hydrogen containers in vessels of this type.

LPG (liquefied petroleum gas, principally propane and butane) which is served by a large contract fleet of smaller vessels of typically 10,000–75,000 DWT. The ships carry either pressure vessels or cryogenic tanks, on some a combination of the two types (pressurised and cryogenic). These vessels can also carry other chemicals that would be classified as an LPG such as ammonia, propylene and vinyl chloride. The LPG trade and the ammonia trade are seasonal and not aligned. The peak LPG trade is in the northern hemisphere autumn/winter and the peak ammonia trade in the northern spring. This seasonal nature affects the carriage cost for both LPG and ammonia. Of particular interest is the carriage of liquid ammonia for the international fertilizer trade in mainly the larger (75,000 DWT) vessels. Ammonia as an intermediate is considered as one of the means for shipping hydrogen around the globe, the LPG shipping fleet could be used for this purpose.

Liquid chemicals are shipped in vessels of about 10,000–40,000 DWT. The vessels are multi-tanked enabling a vessel to carry a range of different cargoes. There are several contract suppliers for chemical transport serving the major industrial chemical markets, i.e. in and around the North Sea, the Mediterranean Sea, the US Gulf and coastal Asia. As well as these regions, the fleet services trans-oceanic shipping across the Atlantic and Pacific oceans. The cost of shipping is dependent on the business cycle. This fleet could be used to ship naphthenes, which are being touted as a means for transporting hydrogen. The small cargo sizes would assist the development of the business.

Clean fuels (naphtha, gasoline, diesel etc.) are used in large quantities, the fleet dedicated to transporting these fuels has larger ships, typically

60,000–120,000 DWT. The fleet is mainly under contract although major oil companies have dedicated ships shipping their products. The fleet could be used to carry naphthenes (which are similar to naphtha) and would benefit from the economy of scale a larger vessel would bring.

Crude oil (and fuel oil) often referred to as dirty cargo is transported by specific ships (fleet). This is because it is costly and somewhat arduous to clean the vessel tanks to facilitate clean fuel transport. Because of the demand for crude oil, these vessels range in size from about 120,000 to over 250,000 DWT. At a larger scale there is an issue with passage through the Suez and Panama canals, the fleet being divided between vessels that can and those that cannot pass through these bottlenecks. Vessels of this scale could be potentially used for the mass transport of hydrogen intermediates, such as naphthenes, and gain from the significant economy of scale offered by very large crude carriers (VLCC).

Contract shipping costs

The cost of movement of liquid fuel cargoes by contract vessels of all shapes and sizes are regularly reported by agencies such as Platts, Petroleum Argus and Waterborne LPG. Prices are marked relative to a standard scale. The two most common scales are the Baltic Dry Index, which is produced by the Baltic Exchange,[2] and Worldscale.[3]

The Baltic Exchanges states that it is:

> "*the world's leading source of independent maritime market data. Our information is used by shipbrokers, owners and operators, traders, financiers and charterers as a reliable and independent view of the dry bulk, tanker, gas and container markets*."

and Worldscale states that:

- *Nominal freight scale applying to the carriage of oil and oil products in bulk by sea.*

---

[2] The Baltic Exchange: www.balticexchange.com

[3] Worldscale Association (London) Ltd & Worldscale Association (NYC): www.worldscale.co.uk

- *An international freight index for tankers that provides a method of calculating the freight applicable to transporting oil by reference to a Standard Vessel (75,000 tonne capacity) on a round trip voyage from one or several load ports to one or several discharge ports.*
- *Includes expenses associated with ports, transit fees, port and voyage time and vessel bunker costs assessed in relation to the Worldscale Standard Vessel basis of calculation in order to produce a comparative nominal freight scale reported in dollars per tonne. Its principle is to provide the same net return per day irrespective of voyage performed for the Worldscale Standard Vessel at WS100.*

However, it should be noted that the actual freight charged is influenced by the specifics of the voyage and the business cycle. Discounts and premiums may be applied as appropriate. Full details of the methodology are described on the Worldscale website.

Build-up of shipping costs

As quotations vary somewhat, it is useful to build up the cost of shipping by an *a priori* method. One such approach was developed by the US Department of Energy (DoE) for estimating the cost of shipping for alternative fuels (such as methanol) on a large scale into the California market.[4] This is adapted here and scaled to present values in Table 6.2.

Four cases are presented, the first two concerning methanol are taken from the DoE paper and the second two on LNG and compressed natural gas (CNG) are developed on the same basis from published data on the capital cost of LNG and CNG vessels.

Referring to the methanol cases, the first is for a typical 40,000-DWT vessel shipping 0.908 PJ (higher heating value [HHV]) of energy as methanol. The second case considers methanol being carried in a VLCC of 250,000 DWT. This transports 5.675PJ (HHV) as methanol. The respective capital costs are M$51.29 and M$144.54. The next sections give the operating costs, fuel usage and cost and port fees. The estimate is

[4] US Department of Energy. (November 1989) *Assessment of Costs and Benefits of Flexible and Alternative Fuel Use in the U.S. Transportation Sector — Technical Report 3: Methanol Production and Transportation Costs*. DOE/PE-0093.

Table 6.2: Estimates for shipping costs on long trans-oceanic voyages for methanol, LNG and CNG.

		Methanol	**Methanol**	**LNG**	**CNG**
Ship capacity	MMscf				345
	m^3			142,500	9,750,362
	DWT	40,000	250,000	65,550	7,476.69
	PJ	0.908	5.675	3,566	0.406
Capital cost	M$	51.29	144.54	266.50	171.43
Operating costs					
Labour	$/day	11,073.72	14,570.68	11,073.72	11,073.72
	M$/y	3.88	5.10	3.88	3.88
Fuel used	t/day	20	30	Boil off	20
	$/t	150	150	150	150
	$/day	3,000	4,500	0	3,000
Port fees/station	$	60,000	80,000	80,000	60,000
Logistics					
Days/year		350	350	350	350
One-way distance	km	6,700	6,700	6,700	6,700
Speed	Knots	12	12	18	18
	km/h	22.224	22.224	33.33	33.33
Sailing time	Days	12.56	12.56	8.37	8.37
Turnaround time	h	24	24	36	18
One-way trips/year		25.81	25.81	35.45	38.36
Sailing days/year		324.19	324.19	296.83	321.23
Port calls/year		25.81	25.81	35.45	38.36
Days in port/year		25.81	25.81	53.17	28.77
ROC (2 y const. 15 y life, 10% DCF)	%	15.19%	15.19%	15.19%	15.19%
Return on capital	$M/y	7.79	21.95	40.47	26.03
Maintenance	%Capex	4%	4%	4%	4%
	M$/y	2.05	5.78	10.66	6.86
Labour	M$/y	3.88	5.10	3.88	3.88
Fuel costs	M$/y	0.97	1.46	0.00	0.96

Table 6.2: (*Continued*)

		Methanol	Methanol	LNG	CNG
Port charges	M$/y	1.55	2.06	2.84	2.30
Insurances	%Opex	15%	15%	15%	15%
	M$/y	1.27	2.16	2.61	1.40
Misc (victualling etc.)	%Opex	10%	10%	10%	10%
	M$/y	0.84	1.44	1.74	2.10
Total OPEX	M$/y	10.56	18.01	21.71	17.50
TOTAL COSTS	M$/y	18.35	39.96	62.18	43.53
Quantity shipped	t/y	516,167.3	3,226,045	1,161,724	143,399
	PJ/y	11.72	73.23	63.89	7.89
Shipping cost	$/t	35.55	12.39	53.53	298.68
	$/GJ (higher heating value [HHV])	1.57	0.55	0.96	5.46

based on a yearly operation of 350 days with a one-way voyage of 6700 km at a speed of 12 knots. The return on capital (ROC) is based on the ship being constructed in 2 years, with a lifetime of 15 years and delivering a 10% discounted cash flow return. To these are added a maintenance charge of 4% of the capital cost per year, with the cost of victualling this gives the total operating costs of M$10.56 and M$18.01 for the two cases respectively. The estimated cost of shipping is $35.55/t or $1.57/GJ (HHV) for the 40,000-DWT, which reduces to $12.39/t or (0.55/GJ [HHV]) for a VLCC, indicating the large economy of scale that can be achieved for transport of large cargoes.

The estimate for the carriage of LNG is developed in a like manner. The capital cost[5] for a 150,000 m^3 vessel is estimated at M$266.5. The carrying volume has been adjusted to allow for the boil-off being used to

[5] Developed from *Oil & Gas Journal* LNG chart quoting Cotton & Co. LNG ship cost of $1200/m^3 2003–2007.

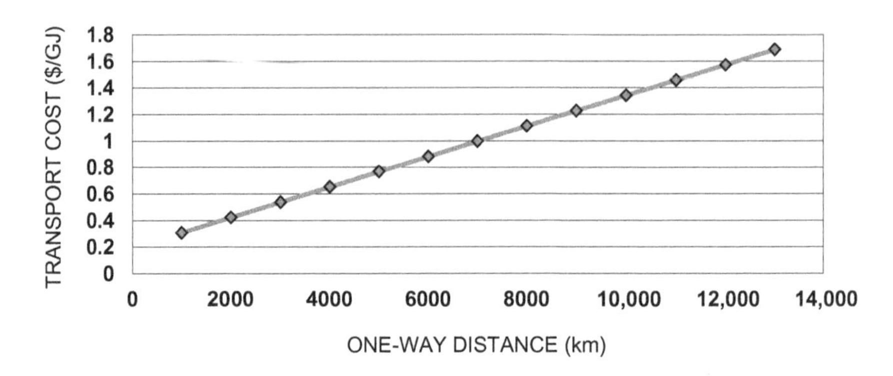

Fig. 6.1: Sensitivity of LNG shipping cost to one-way distance.

fuel the ship. Modern LNG ship speeds are 18 knots; this reduces the sailing time between long-distance destinations. Following the method for methanol ships, the estimated cost of carriage for LNG is $53.5/t or $0.96/GJ (HHV), which is typical of reported costs for long oceanic voyages.[6] The cost of shipping is a function of the distance travelled, the sensitivity of the LNG shipping cost to one-way distance, as illustrated in Figure 6.1.

The figure shows that for a 6700-km journey (e.g. Australia NWS to Japan) the shipping cost of LNG is about $1/GJ but would be double this for shipping distances of over 17,000 km (Australian NWS to United Kingdom).

Compressed natural gas has been touted as an alternative to LNG thus saving the large capital cost of the liquefaction plant.[7] The capital cost of the ship based on a Coselle system is estimated at M$171.0. The estimate method follows that for LNG. Insurance is lower for this option. The estimated cost of gas transport is $298.7/t or $5.38/GJ (HHV), which is considerably higher than the estimated for LNG for such a long voyage. This system could be potentially used for hydrogen transport saving the larger hydrogen liquefaction costs and is discussed more fully later in this chapter.

[6] Timera Energy: https://timera-energy.com/deconstructing-lng-shipping-costs/

[7] Wagner JV, van Wagensveld S. (8–10 October 2002) Marine transportation of compressed natural gas: A viable alternative to pipeline or LNG. In: *The 2002 SPE Asia Pacific Oil and Gas Conf and Exhibition*, Melbourne, Australia (8–10 October 2002), SPE Technical Paper 77925.

Cost of Shipping Liquid Hydrogen

We are concerned with the cost of shipping hydrogen as liquid over trans-oceanic distances to connect the areas where hydrogen could be cheaply produced to distant markets. To assist in identifying the hurdles and cost for this, we consider the supply of hydrogen from the north-west of Australia (a major area for the production of LNG) to Tokyo Bay in Japan, which is potentially a major market for hydrogen and is 3622 nautical miles or 6700 km distant.

We consider the logistics to comprise:

- Production of hydrogen (method not defined),
- Liquefaction of the hydrogen and local storage,
- Transfer to large trans-oceanic vessel of similar dimension to an LNG vessel,
- Unloading and regasification at the receiving port.

Hydrogen liquefaction

For trans-oceanic supply of hydrogen to distant markets we are concerned with the mass production of liquid hydrogen. We may take as a guide the production cost of LNG and apply this to hydrogen produced from natural gas by steam reforming/gasification. Analysis along these lines results in very high hydrogen liquefaction costs (e.g. over $4000/t). This is principally because the very high capital cost of LNG plant are usually constructed in remote and difficult terrain. For example a 6 Mt/y LNG operation would typically expend over M$14000. It is not clear if these published expenditures encompass the development of the supplying gas fields and the LNG shipping fleet.[8] A further problem is that to liquefy hydrogen it takes about 10 times the refrigeration power than that required to liquefy the same mass of natural gas.

A recent study by Connelly *et al.*[9] gives an indication of the cost of liquefaction plant. The process plant modelled is shown in Figure 6.2.

[8]Yost C, DiNapoli R. (2003) Benchmarking study compares LNG plant costs. *Oil Gas J* **101**: 56–59.

[9]Connelly E, Penev M, Elgowainy A, Hunter C, (9 September 2019) *US DOE Hydrogen and Fuel Cells Program Record, No. 19001, Current Status of Hydrogen Liquefaction Costs.*

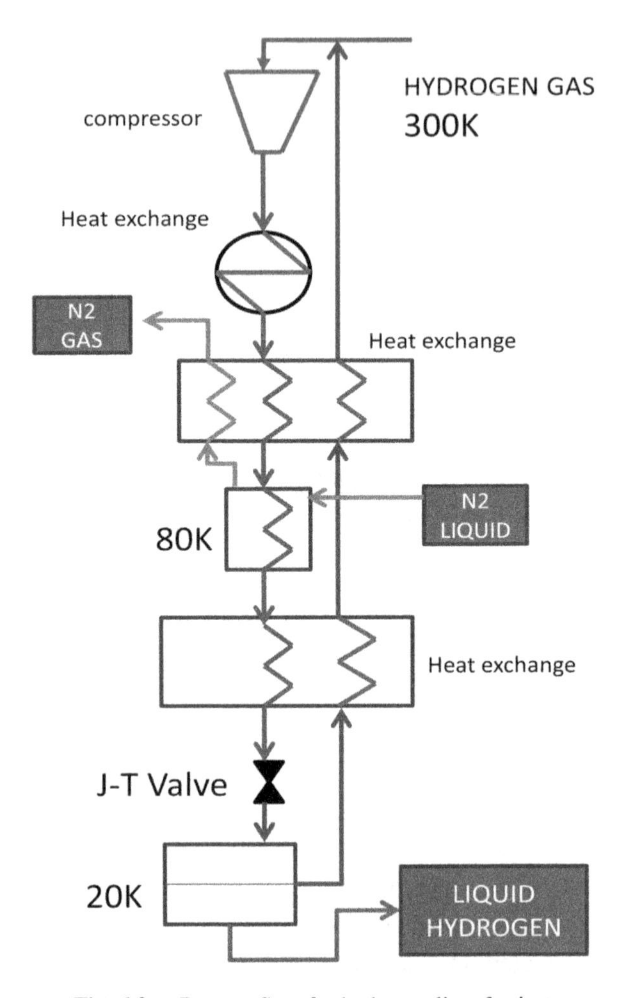

Fig. 6.2: Process flow for hydrogen liquefaction.

In the process modelled, hydrogen gas at ambient temperature and 2 MPa pressure from a steam methane reformer is further compressed, passed through heat exchangers, which includes a liquid nitrogen bath with the duty to reduce the hydrogen temperature to 80 K, well below the J-T inversion temperature. Additional heat exchange reduces the hydrogen temperature further before expansion through a J-T valve to produce liquid hydrogen.

The ideal energy consumption for this process is 2.88 kWh/kg of liquid hydrogen, but in practice the typical energy consumption is in the range of 10–20 kWh/kg.

At the time of writing, the problem with this approach was that the current hydrogen liquefaction plants are relatively small, typically 6–70 t/d. Although these show good economy of scale it is not clear if there are upper bounds to the scaling of the process plant. The economy of scale reported by Connelly to 200 t/d capacity was for a California location:

$$\textbf{Capital Cost (M\$)} = \textbf{5.6} \times \textbf{(Hydrogen Capacity [t/d])}^{\textbf{0.8}} \quad \textbf{(6.1)}$$

Using this as a basis and assuming that a large-scale plant would comprise multiple units of 200-t/d capacity, the estimated cost of liquefaction for a central type facility (1100 t/d) is developed in Table 6.3.

Table 6.3: Estimated cost of hydrogen liquefaction (US Gulf).

Hydrogen production rate	t/d	1100
Capital cost (California)	M$	2755
Location factor (California to US Gulf)		0.862
CAPEX 2018 US Gulf	M$	2611.5
ROC (2 y const. 20 y life 10% DCF)		13.84%
Return on capital	M$/y	361.3
Operating Costs	%Capex	
Labour	1%	26.12
Maintenance	3.00%	78.35
Catalyst and chemicals	1%	26.12
Insurance etc.	1.50%	39.17
Total		169.75
Power	kWh/kg	10
	MWh/y	3,739,400
	$/MWh	50

(Continued)

Table 6.3: *(Continued)*

	M$/y	186.97
TOTAL COSTS	M$/y	718.11
	$/t	1920.40
	$/GJ (lower heating value [LHV])	16.02

The hydrogen production rate of 1100 t/d is delivered by five plants of 200 t/d capacity and one of 100 t/d capacity at a capital cost for a Californian location of M$2,755. This is translated to a US Gulf location using a location factor of 0.862 to give a capital cost of M$2,611. The return on capital for a 2-year construction, 20-year life and delivering a 10% discounted cash flow return on the capital expended is M$361.3/y. Operating costs (labour, maintenance etc.) is M$169.75/y.

The power consumption is assumed to be 10 kWh/kg, which is at the low end of the range reported and the power costs are assumed to be $50/MWh. This delivers a hydrogen production cost of $1920.0/t or 16.02/GJ (lower heating value [LHV]).

In Chapter 5 the work of Amos[10] was reported on the cost of liquefaction and storage of hydrogen. This estimate on similar statistics gave an estimate for the liquefaction of hydrogen including storage vessels of about $1000/t, which is about half the value based on this more recent DoE study. This can be substantially explained by the much lower capital cost in the Amos work and it is a moot point if that study took into account fully the hydrogen liquefaction process plant (including a nitrogen liquefaction unit) into account.

The cost of liquefaction is very sensitive to the process efficiency (specific power usage) and the cost of power. The sensitivity to the process efficiency is illustrated in Figure 6.3.

The ideal power usage is 2.88 kWh/kg, which would produce a liquefaction cost below $1,600/t. It can be seen that power usage above 12 kWh/kg (MWh/t) would have liquefaction costs above $2000/t, all other parameters being equal.

[10]Amos WA. (November 1998) *Cost of Storing and Transporting Hydrogen.* NREL/TP-570-25106.

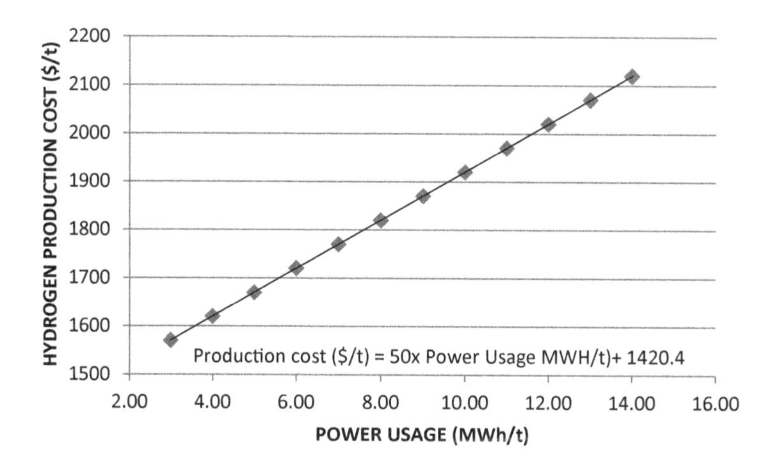

Fig. 6.3: Sensitivity of hydrogen liquefaction cost to power usage.

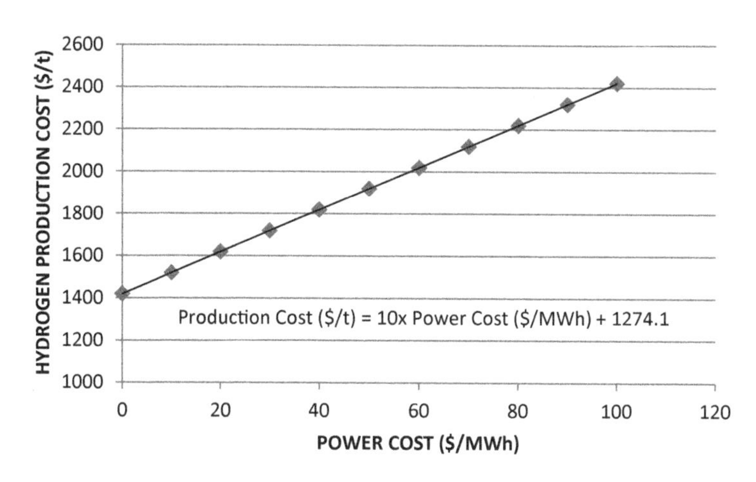

Fig. 6.4: Sensitivity of hydrogen liquefaction costs to the cost of power.

The sensitivity of hydrogen liquefaction costs to the cost of power is illustrated in Figure 6.4. As can be seen at low power costs ($20/MWh) the cost of hydrogen liquefaction can be below $1600/t, all other variables being equal. Very low power costs such as this are associated with regions that can continuously generate power from hydropower schemes. Most low-cost renewable power is in the $50/MWh region, but if this is not available it can be seen that if power cost is above $60/MWh the cost of hydrogen liquefaction rises to over $2000/t.

Shipping liquid hydrogen

To develop an estimate for the cost of shipping large volumes of liquid hydrogen, it is assumed that ships similar in size and cost to those being used in the LNG trade could be used. The method for developing the estimate has been given in Table 6.2 and is developed for liquid hydrogen in Table 6.4. For comparison the case for LNG is included. Note the difference that the final costs estimate in energy terms is on a LHV basis for both hydrogen and LNG.

Table 6.4: Estimate for shipping liquid hydrogen compared to LNG.

		LNG	**Liquid H_2**
Ship capacity	m^3	142,500	142,500
	DWT	65,550	10,096
	GJ	3,565,806	1,431,545
Capital cost	MM$	266.50	266.50
Operating costs (Opex)			
Labour	$/day	11,073.72	11,073.72
	M$/y	3.88	3.88
Fuel	t/day	Boil off	Boil off
	$/t	150	150
Port fees/station	$	80,000	80,000
Logistics			
Days/year		350	350
One-way distance	km	6700	6700
Speed	knots	18	18
	km/h	33.33	33.33
Sailing time	days	8.37	8.37
Turnaround time	h	36	36
One-way trips/year		35.45	35.45
Sailing days/year		296.83	296.83
Port calls/year		35.45	35.45
Days in port/year		53.17	53.17
ROC (2 y cost, 15 y life 10% DCF)	%	15.19%	15.19%

Table 6.4: (*Continued*)

		LNG	Liquid H_2
Return on capital	$M/y	40.47	40.47
Maintenance	%Capex	4%	4%
	M$/y	10.66	10.66
Labour	M$/y	3.88	3.88
Fuel costs	M$/y	0.00	0.00
Port charges	M$/y	2.84	2.84
Insurances	%Opex	15%	15%
	M$/y	2.61	2.61
Misc (victualling etc.)	%Opex	10%	10%
	M$/y	1.74	1.74
Total OPEX	M$/y	21.71	21.71
TOTAL COSTS	M$/y	62.18	62.18
Quantity shipped	t/y	1,161,724	178,930
	PJ/y	63.89	25.37
Shipping cost	$/t	53.53	347.52
	$/GJ (lower heating value [LHV])	1.07	2.91

The estimate for shipping of liquid hydrogen is $347/t or $2.91/GJ (LHV); this compares to $55.5/t or $1.07/GJ (LHV) for LNG shipping.

The analysis indicates that each ship will be able to move about 180,000 t/y at this optimum. This would indicate that a central facility would need at least two, or possibly three, ships for the carriage of hydrogen from a central facility–sized system to a destination at a distance of 6700 km.

Regasification of Liquid Hydrogen

After delivery at the receiving terminal the liquid hydrogen will need to be stored and regasified prior to use. Storage has been discussed in

Chapter 5. Liquid hydrogen can be regasified similar to LNG regasification, which is described below.

One of the earliest methods, still in use, is the open rack vaporiser. In this method LNG is passed through a network of pipes over which is sprayed seawater. The seawater at ambient temperature warms the pipework, which transfers the heat to the liquid LNG, which returns to the gaseous state. The cooled seawater falls into a catchment pond and is ultimately discharged back to the sea. This is a relatively simple and cheap regasification method. The main operating cost is the cost of pumping the seawater. The main operating problem is that cold water returning to the sea may have environmental impacts.

A more compact method is that of the submerged combustion vaporiser, as shown in Figure 6.5. In this method the LNG to be gasified is passed through a pipeline system held within a water bath. The bath is heated by means of a combustion chamber, which is substantially under the water in the bath. The combustion chamber is fuelled by a fuel gas (or some of the natural gas) and air. The combustion chamber heats the water in the bath and the heat is subsequently transferred to the pipework regasifying the LNG. Flue gases from the combustion sparge into the tank and travel up through the water ensuring high efficiency in delivering the combustion heat to the water.

The advantage of this method is that it is quite compact. It can be built within a ship hull in a floating storage and regasification unit (FSRU) operation where a trans-oceanic tanker discharges its cargo into the FSRU

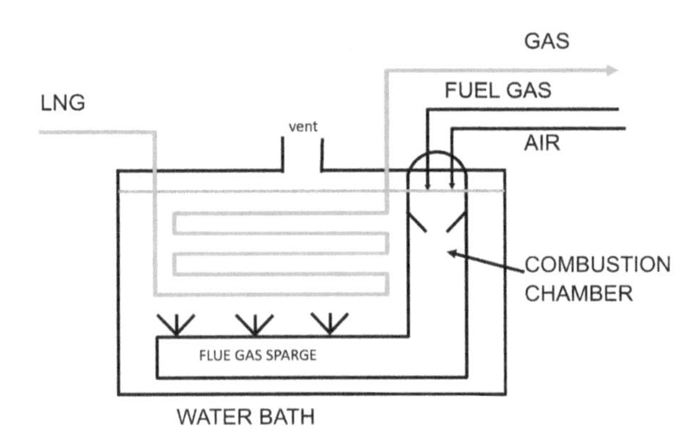

Fig. 6.5: Submerged combustion vaporiser.

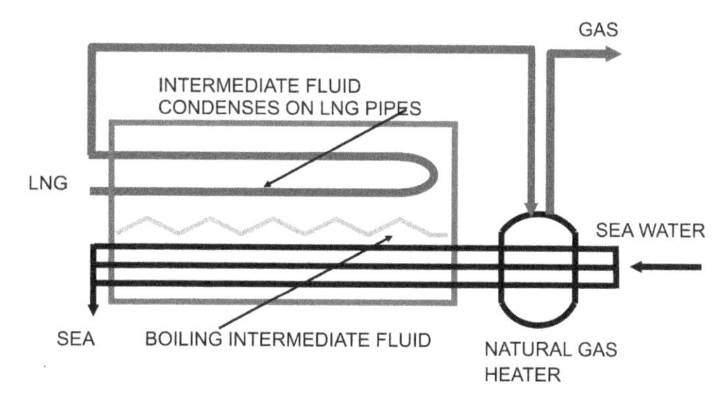

Fig. 6.6: Intermediate fluid vaporiser.

(see later). A disadvantage of this method is that pollutants such as nitrogen oxides (NOx) are ultimately discharged to atmosphere though the water bath.

Another method is the intermediate fluid vaporiser. This is illustrated in Figure 6.6. A pipeline network with the duty of vaporising the LNG is held near the top of a pressure vessel. The bottom of the vessel contains a fluid that acts as a heat transfer agent, typically propane. The intermediate fluid is boiled by seawater, which is heated as required by a natural gas heater and then discharged back to the sea. The gasified LNG stream is also passed to the heater to finalise the evaporation and bring the gas to near ambient temperature. The boiling fluid condenses on the LNG pipework transferring heat and vaporising the liquid to gas.

This system is quite compact and similar in design to large process heat exchangers. They are very efficient and are often used in FSRU operations.

Adding value in regasification facilities

One interesting point of note is that in many regasification facilities is extensive value extracted from the refrigeration power being delivered as LNG (and potentially liquid hydrogen). Because LNG is delivered at −165°C the regasification process can be used to facilitate operations that require refrigeration. This practice is common in Japanese regasification facilities. The LNG goes from the delivery vessel to a land-based storage tank from where it can be fed for refrigeration use, prior to final regasification. There are several options, e.g.:

- Integration with air separation units to reduce compressor power costs,
- Manufacture of dry ice (solid carbon dioxide),
- Use of the volumetric expansion of the liquid to gas transition for power generation,
- Food processing — food freezing.

This is applicable to land-based regasification facilities but it could be difficult to utilise these schemes in FSRU operations.

Regasification cost

The regasification cost of LNG is well known and reported in trade journals (e.g. *Argus*) for various locations. LNG regasification capital costs are discussed in *IGU World Gas LNG* reports.[11] An extensive review of the costs is given by Uemura and Ishigami.[12] The *IGU World 2018* report suggests a capital cost of \$285/t for land-based regasification terminals and \$158/t for floating storage and regasification units.

There are no large-scale hydrogen regasification units. It is not clear how these would materially differ from LNG regasification and on what basis the LNG data could be used as guide for estimating the cost for regasification of liquid hydrogen. The heat required to regasify hydrogen is approximately 8.73 kJ/mol on a theoretical basis compared to 5.66 kJ/mol for methane (the main component of LNG). However, on a mass basis this is 4.33 kJ/kg and 0.511 kJ/kg respectively. In this estimate we have assumed that the overall process is best compared on the specific energy requirement basis, i.e. for a similarly constructed and sized plant, the throughput for hydrogen would be only about 9% (by mass) of that for LNG regasification of a similar-sized facility.

---

[11] *International Gas Union*: www.igu.org

[12] ERIA. (2018) Investment in LNG supply chain infrastructure estimation. In: Uemura T, Ishigami K, (eds), *Formulating Policy Options for Promoting Natural Gas Utilization in the East Asia Summit Region Volume II: Supply Side Analysis. ERIA Research Project Report 2016-07b*. ERIA, Jakarta, Indonesia, pp. 67–80.

Using the methods previously described, the statistics for shore-based LNG, FSRU and hydrogen are detailed in Table 6.5.

The statistics are based on a 3,300 kt/y LNG regasification plant using the *IGU* estimate for capital cost. All process plant is assumed to be

Table 6.5: Estimate for regasification of shore based LNG, LNG-FSRU and hydrogen.

		LNG	**FSRU**	**HYDROGEN**
Nominal capacity	kt/y	3300	3300	298.05
t/d	kt/d	9.71	9.71	0.88
PJ/y		179.51	179.51	35.47
Capital recovery (2 y const, 20 y life, 10% DCF)	%	13.57%	13.57%	13.57%
Base capital cost	M$	940.50	521.40	
Base year		2018	2018	
Location factor		1.000	1.000	
NF index		1.0000	1.0000	
FINAL CAPEX	M$	940.50	521.40	940.50
Capital Charges	M$/y	127.59	70.74	127.59
Working capital (%Capex)	M$	96.09	96.09	78.90
Return of working capital	M$/y	9.61	9.61	7.89
CAPITAL CHARGES (C)		137.20	80.35	135.48
<u>Operating Costs</u>				
Labour (1.44% Capex/y)	M$/y	13.54	7.51	13.54
Maintenance (1.73% Capex/y)	M$/y	16.27	9.02	16.27
Insurance (1.50% Capex/y)	M$/y	14.11	7.82	14.11
Catalysts & Chemicals	M$/y	1.00	1.00	1.00
NON-FEED OPEX (O)		59.03	33.17	59.03
<u>Heating gas usage</u>	kt/y	36.04	36.04	12.81
Natural gas at $8/GJ; hydrogen at $3000/t	$/t	440.00	440.00	3000.00
Gas reheating costs	M$/y	15.86	15.86	38.43
TOTAL NON-FEED COSTS	M$/y	197.98	121.55	218.84
GAS OUTPUT	PJ/y	177.53	177.53	33.94
REGASIFICATION COSTS	$/t	60.22	36.97	767.35
High heating value [HHV] for LNG, low heating value [LHV] for hydrogen	$/GJ	1.12	0.68	6.45

constructed in 2 years, has a 20-year life and delivers a 10% DCF (discounted cash flow) on capital employed. The terminal stores product for 30 days. This requires an assessment of the working capital for the storage which is based on $330/t for LNG and $3000/t for hydrogen. The return required on the working capital is 10%.

The operation requires energy to regasify the cryogenic liquids. This has been estimated by assuming that the product is used for this role and that the thermal efficiency of the process is 85%. Natural gas is assumed to be charged at $8/GJ ($140/t) and hydrogen at $3000/t. The reheating cost for hydrogen is more than double that for LNG.

The estimate for a new shore-based LNG operation is that the regasification cost is about $60/t ($1.12/GJ [HHV]) and for a FSRU of the same capacity $37/t (0.68/GJ [HHV]). These are more than the typically reported values but are for new plant and conservative values for the heating costs; for instance, existing facilities have sunk capital and may use cheaper energy sources such as fuel oil.

For the regasification of liquid hydrogen on a shore-based facility is estimated at $767/t ($6.45/GJ [LHV]). This is much higher than the cost for LNG regasification.

Summary

The costs of shipping hydrogen as liquid over a trans-oceanic voyage is summarised as:

- Liquefaction and storage (Table 6.3) $1920/t or $16.02/GJ (LHV)
- Shipping (Table 6.4) $347/t or $2.91/GJ (LHV)
- Regasification (Table 6.5) 767/t or 6.45/GJ (LHV)
- TOTAL $3034/t or $25/38/GJ (LHV)

As can be readily seen from this, the high cost of liquefaction shipping and regasification has led to proposals for alternative approaches for delivering hydrogen to long-distant markets.

Delivering Hydrogen as Compressed Gas

The delivery of natural gas as CNG to distant trans-oceanic markets has been considered as an alternative to LNG. Hydrogen could be delivered in a similar manner. Delivering hydrogen as a compressed gas aims to:

- Deliver hydrogen without the need for the high cost of liquefaction,
- Eliminates storage costs (the ship is used as the store),
- Lowers the hydrogen plant cost since purification for cryogenic purposes is not required,
- Eliminates regasification costs.

For compressed gas delivery the economics are dominated by the cost of shipping (typically about 90% of the costs). There are currently three approaches to the construction of CNG vessels:

- Coselle[13]: In this system a long (10 km) small-diameter (6″) pipe is coiled into a coselle. The coselle has a capacity of 3.2 MMscf. A 60,000-DWT tanker can carry 108 coselles transporting about 345 MMscf (9.75 Mm^3) of gas.
- Votrans[14]: This is a variant in which larger diameter pipes are held in insulated cold storage units. Carrying gas a lower temperature allows for lower pressure operation.
- Gas Transport Modules (GTM): In this system reinforced pressure vessels made from high-strength alloy (HSLA) steel pipe wrapped with composites. These are approximately 20% lighter than steel tanks. This is a variation on the tanks in use for the storage of hydrogen in vehicles, though on a much larger scale.

---

[13] Wartsila Encyclopedia of Marine Technology: www.wartsila.com, www.coselle.com and www.gev.com

[14] Enersea Transport Inc. www.enersea.com

The economics of the proposal is developed after Mokhatab and Economides[15] for CNG operation between Qatar and Jamnagar (Gujarat, India). The data in Table 6.6 has been escalated to 2018 values and developed for a 6700-km sea voyage in the same manner as that developed for the transport of LNG and liquid hydrogen reported in Table 6.4. For the same volume available on a ship much less hydrogen by weight is able to be transported, which results in higher specific shipping cost for hydrogen. The resulting estimate is that the carriage cost of CNG is \$303/t or

Table 6.6: Economics of shipping compressed natural gas and hydrogen over long distances.

		CNG	Hydrogen
Ship capacity	MMscf	345	345
	cm	9,750,362	9,750,362
	DWT	7476.69	825.24
	GJ	406,718.95	117,012.32
Capital cost	M\$	171.43	171.43
Operating costs			
Labour	\$/day	11,073.72	11,073.72
	M\$/y	3.88	3.88
Fuel	t/day	20	20
	\$/t	150	150
	\$/day	3000	3000
Port fees/station	\$	60,000	60,000
Logistics			
Days/year		350	350
One-way distance	km	6700	6700
Speed	knots	18	18
	km/h	33.336	33.336
Sailing time	days	8.37	8.37
Turnaround time	h	18	18
One-way trips/year		38.36	38.36

[15] Mokhatab S, Economides MJ. (January 2007) Compressed natural gas — Another solution to monetize stranded gas. *Hydrocarbon Processing*, p. 59.

Table 6.6: (*Continued*)

		CNG	Hydrogen
Sailing days/year		321.23	321.23
Port calls/year		38.36	38.36
Days in port/year		28.77	28.77
Return on capital (2 y const, 15 y life, 10% DCF)	%	15.19	15.19
Capital costs		26.03	26.03
Maintenance	%Capex	4	4
	M$/y	6.86	6.86
Labour	M$/y	3.88	3.88
Fuel costs	M$/y	0.96	0.96
Port charges	M$/y	2.30	2.30
Insurances	%Opex	15	15
	M$/y	2.10	2.10
Misc (victualing etc)	%Opex	10	10
	M$/y	1.40	1.40
Total OPEX	M$/y	16.80	16.80
TOTAL COSTS	M$/y	43.53	43.53
Quantity shipped	t/y	143,399	15,827
	PJ/y	7.89	2.24
Shipping cost	$/t	303.56	2750.22
	$/GJ	5.46	22.93

$5.46/GJ (HHV) and for hydrogen $2750/t or $22.9/GJ (LHV). These shipping costs are much higher than those estimated for LNG and liquid hydrogen and indicate that carriage as compressed gas is economically unsuitable for long trans-oceanic trade. This is generally well known for CNG and the main thrust of efforts to develop CNG has been directed to shorter shipping distances. The sensitivity of the shipping cost over short distances is illustrated in Figure 6.7. This illustrates the potential lower cost of shipping compressed gas, and possibly compressed hydrogen, over distances of less than 2000 km.

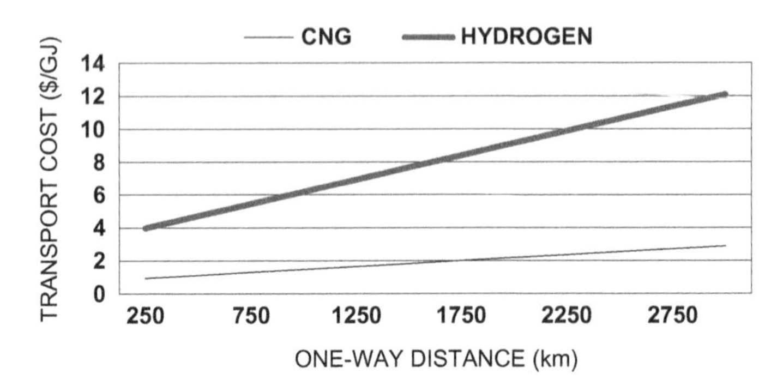

Fig. 6.7: Sensitivity of shipping cost of shipping CNG and hydrogen over short distances (CNG in HHV, hydrogen in LHV).

Transport of Hydrogen as an Intermediate

The difficulty and high cost of delivering hydrogen across trans-oceanic voyages has spurred interest in alternative methods for achieving this end. One way is to consider moving the hydrogen via an intermediate compound. For this method the chosen intermediate compound must:

- Be able to contain a significant quantity of hydrogen,
- Be easy to manufacture at the point where hydrogen is available,
- Be easily shipped,
- Be easily decomposed at the receiving end to liberate the hydrogen.

From these requirements two alternatives have emerged: naphthenes and ammonia.

Transport of Hydrogen as a Naphthene (The Spera *Process)*

Naphthene is a generic industry term for *cyclo*-alkane and a great many naphthenes are found in crude oil. Of most interest are naphthenes (*cyclo*-alkanes) with more than six carbon atoms. For these compounds, using an appropriate catalyst, the compound can be converted into an aromatic compound, thereby releasing three hydrogen molecules. The reaction is reversible. *Cyclo*-alkanes containing 10 carbon atoms in the rings can also be considered (e.g. decalin). These compounds could carry more

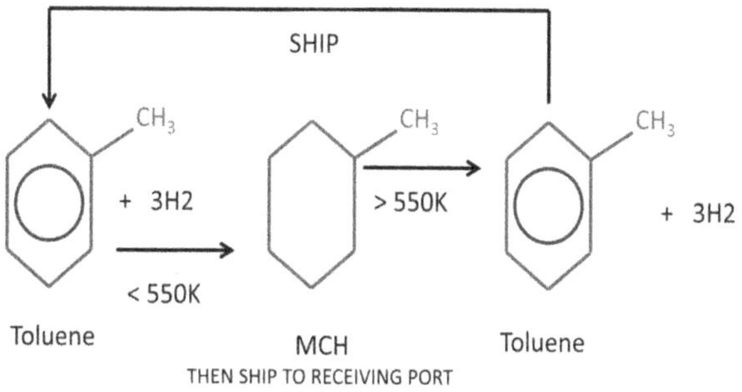

Fig. 6.8: *Spera* process.

hydrogen but the resulting product C_{10} aromatic, e.g. naphthalene is solid at room temperature, whereas single-ring compounds (with some exceptions) are, at ambient temperature, liquid in both the *cyclo*-alkane and aromatic states.

Of particular interest is methyl-*cyclo*-hexane (MCH) and toluene which is the basis for the *Spera* process promoted by the Chiyoda Corporation.[16] MCH contains 6% by weight available hydrogen, which is used as a carrier molecule. This system is illustrated in Figure 6.8.

In this process:

- Hydrogen is produced by any means at a remote location,
- The hydrogen is used to reduce toluene to MCH,
- The MCH is shipped to a distant market,
- The MCH is dehydrogenated to toluene releasing hydrogen,
- The toluene is shipped back to the hydrogen production site for reuse.

Rather than toluene/MCH it would be feasible to use a benzene/*cyclo*-hexane system. Reduction of benzene to *cyclo*-hexane is widely practiced in the petrochemical industry where it is an integral process for the production of nylon. This would have the advantage of carrying more hydrogen, *cyclo*-hexane contains over 7% available hydrogen, but it has the disadvantage of requiring the production and shipping of benzene, which is highly toxic.

[16] Chiyoda Corporation. *The Spera Hydrogen System*. www.chiyodacorp.com

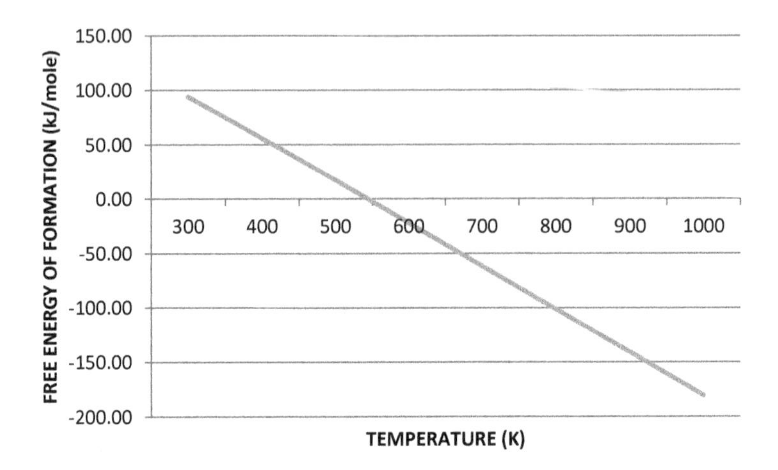

Fig. 6.9: Free energy change for conversion of MCH to toluene.

Process thermodynamics

The detail of the process for the conversion of naphthenes into aromatics is governed by the thermodynamics of the reaction.

$$\mathbf{MCH \leftrightarrow Toluene + 3H_2} \tag{6.2}$$

The overall free energy change for the process is illustrated in Figure 6.9.

For a reaction to proceed with a practical product yield, the change in free energy has to be close to and preferably lower than zero. This graph illustrates that for conversion of MCH into toluene and liberating hydrogen the practical temperature has to be over about 550 K. Conversely the conversion of toluene into MCH would be best performed below 550 K. The equilibrium is influenced by the pressure of the system. The conversion of toluene to MCH will benefit from high hydrogen partial pressure and conversely the conversion of MCH into toluene from low hydrogen partial pressure.

Furthermore, for toluene into MCH the reaction will release 205 kJ/mol, i.e. at the hydrogen production end, and will require this amount of energy to be provided at the receiving end.

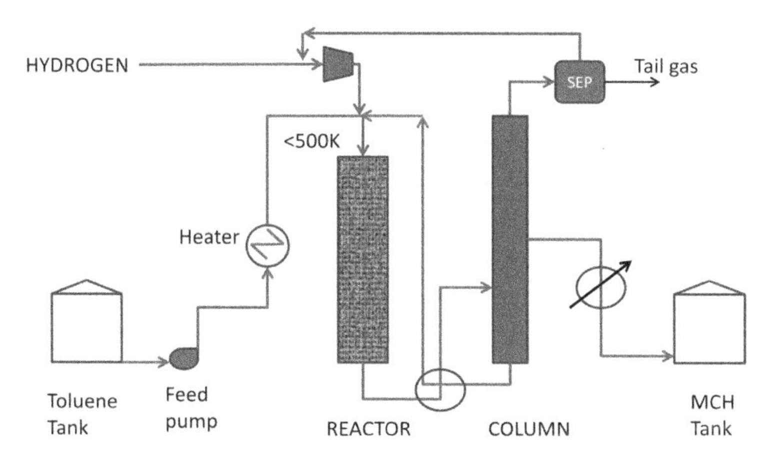

Fig. 6.10: Basic process flow for the hydrogenation of toluene into MCH.

Process flow for toluene to MCH

A possible basic flow for the process is illustrated in Figure 6.10. The process is very similar to the conversion of benzene into *cyclo*-hexane, which is a process of great importance to the petrochemical industry and widely described in the literature.[17] Toluene is fed from a storage tank and heated to the reaction temperature prior to being fed to the top of the reactor. Typically this is less than 200°C, where it is mixed with compressed hydrogen (typically about 3.5 MPa). The reactor is charged with catalyst (nickel supported on alumina is common) where conversion of the toluene into MCH occurs. The products exiting the reactor are fed to a distillation column, which separates unreacted toluene (b.p. 110°C) from the MCH (b.p. 101°C) and excess hydrogen and tail gases, the latter from spurious cracking reactions. A separation system (e.g. membrane) separates excess hydrogen from the tail gas and is recycled. The MCH flows to a storage tank for shipping. There are potentially many variations on this, e.g. the use of solvent extraction techniques to separate the aromatics from the naphthenes.

[17] For example, see Sanderson JR, Renken TL, McKinney MW. (June 15, 2004) *US Patent 6,750,374 to Huntsman Petrochemical Corporation.*

As noted above, this part of the process is exothermic (generates heat) and as a consequence heat exchange is required to remove the heat energy.

Process flow for MCH to toluene

The possible process flow for the conversion of MCH into toluene, thereby liberating the hydrogen at the recipient's end of the chain, is illustrated in Figure 6.11. As noted above this reaction requires a higher temperature (more than 550 K) and is endothermic. The reaction is similar to the refinery operation of naphtha reforming to produce high-octane (high-aromatic) gasoline for which there is considerable literature.

In a basic process MCH would be fed via a heater to the top of the reactor. The feed would be heated to typically 500°C (over 750 K) to allow for the reaction heat required by the reaction. At this temperature there would be more cracking from side reactions, which would result in higher levels of tail gas being produced. To suppress this a small amount of hydrogen would be mixed with the MCH feed.

The reactants entering the top of the reactor would pass over a catalyst, which would typically be platinum supported on alumina promoted by rhenium (naphtha reforming catalyst). The conversion of a pure feed such as MCH would be less severe than for naphtha reforming, nevertheless, the process would be broadly similar. The endothermic nature of the reaction would rapidly cool the feed and unwanted cracking reactions

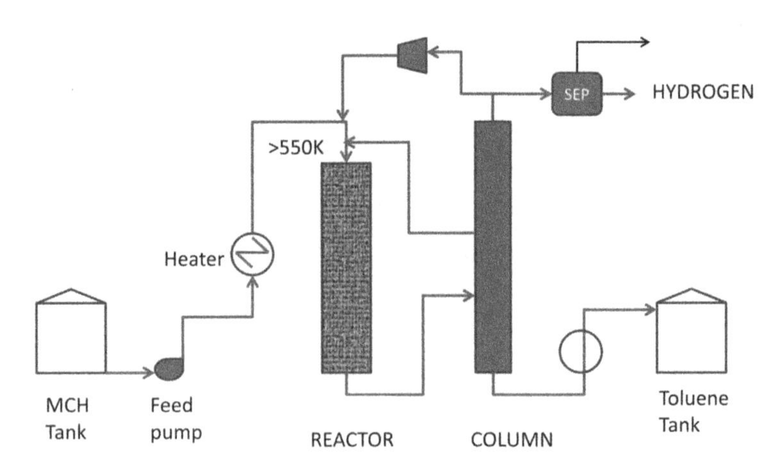

Fig. 6.11: Basic process flow for the conversion of MCH into toluene and hydrogen.

would foul the catalyst, which would require regeneration. The process of naphtha reforming teaches that there are many approaches to optimising this process such as multiple beds, fluid beds etc.

The products would leave the base of the reactor and pass to a distillation column to separate unconverted MCH from the toluene, hydrogen and the cracked gases. The toluene would be passed to a tank ready for re-shipping to the MCH facility. Unconverted MCH would be recycled. The product hydrogen would be separated from the gas stream with a small portion being passed back to the reactor.

Process Economics

The process economics for both the toluene to MCH step and the MCH to toluene step are developed in Table 6.7 and are for a central facility with 373 kt/y hydrogen production. The economics is based on the economics of a naphtha reforming given by Maples[18] for a reformer with a capacity of 30,000 bbl/d; other descriptions of naphtha reformer technology and economics are also available.[19]

In analysing this method of hydrogen transport one of the critical factors is to take fully into account the process efficiencies of the toluene to MCH step and the MCH to toluene step. Since toluene is the effective hydrogen carrier, losses of the toluene at each end of the chain will be required to be made up with fresh toluene.

The first step of toluene to MCH is likely to be highly efficient since it is a well-known low-temperature process. If this process can be delivered at 99% pass through selectivity then, allowing for some recycle an overall efficiency of over 98% may be possible. For this analysis it is assumed the losses are 1.5% (98.5% overall selectivity) with all of the losses as tail gas.

However, for the second step of MCH to toluene significant losses are to be expected as a consequence of the higher operating temperatures, the acidic nature of catalyst supports and losses associated with intermediary reheating of the feed as it cools due to the endothermic nature of the

[18] Maples RE. (19993) *Petroleum Refinery Process Economics*. PennWell Books, Tulsa, OK.

[19] For example *Oil & Gas Journal*, Refinery Process Handbook.

Table 6.7: Economics for the conversion toluene to MCH and MCH to toluene.

Toluene to MCH			**MCH to Toluene**		
Toluene density	g/L	870	MCH density	g/L	770
Hydrogen	kt/y	373	Hydrogen	kt/y	373
Toluene required	kt/y	5728.04	MCH required	kt/y	6092.33
	bbl/d	121805.0		bbl/d	146,376.56
Reformer basis	bbl/d	30000	Reformer basis	bbl/d	30,000
Cost 1991	M$	37.5	Cost 1991	M$	37.5
Cost 2018	M$	85.53	Cost 2018	M$	$ 85.53
Capacity increase ($n = 0.8$)	M$	262.39	Capacity ($n = 0.8$)	M$	$ 303.94
ROC (3 y const, 20 y life, 10% DCF)	%Capex	14.60	ROC (3, 10, 20)	%Capex	14.60
	M$/y	$ 38.32		M$/y	$ 44.38
OPEX	%Capex		OPEX	%Capex	
Labour	1%	$ 2.62	Labour	1%	$ 3.04
Maintenance	3.50%	$ 9.18	Maintenance	3.50%	$ 10.64
Insurance	1.50%	$ 3.94	Insurance	1.50%	$ 4.56
Catalysts and chemicals	1%	$ 2.62	Catalysts and Chemicals	1%	$ 3.04
Subtotal	M$/y	$ 18.37	Subtotal	M$/y	$ 21.28
Electricity required	kWh/bbl	1	Electricity	kWh/bbl	1
	MWh/y	41,413.70		MWh/y	49,768.03
	$/MWh	50		$/MWh	50
	M$/y	$ 2.07		M$/y	$ 2.49
Fuel required	BTU/bbl	300,000	Fuel required	BTU/bbl	300,000
	GJ/bbl	0.317		GJ/bbl	0.317
	PJ/y	13.11		PJ/y	15.753
Tail gas	PJ/y	4.30	Tail gas	PJ/y	13.60
Net fuel required	PJ/y	8.8	Net fuel required	PJ/y	2.15
Fuel cost	$/GJ	6	Fuel cost	$/GJ	6
TOTAL Non-feed OPEX	M$/y	$ 111.63	TOTAL NF OPEX	M$/y	$ 81.04
Toluene makeup	t/y	85,920.55	Toluene makeup	t/y	272,081.74
Toluene price	$/t	800	Toluene price	$/t	800
	M$/y	$ 68.74		M$/y	$ 217.67

Table 6.7: (*Continued*)

Toluene to MCH			**MCH to Toluene**		
Hydrogen costs (SMR/CCS)	\$/t	1151.0			
Process efficiency		98%			
Hydrogen required	kt/y	380.6			
Hydrogen costs	M\$/y	438.1			
TOTAL COSTS	M\$/y	\$ 618.45	TOTAL COSTS	M\$/y	\$ 298.70
MCH COST	\$/t	\$ 101.4			
H_2 EQUIVALENT COST	\$/t	\$ 1658.1	H_2 COST	\$/t	\$ 800.82

reaction. Analysis of reformer data[20] would indicate a loss of up to 5% or 6% of the feed. Because the MCH dehydrogenation is expected to be less severe than naphtha reforming, in this analysis it is assumed that the loss in this second step is 4.5% overall, i.e. allowing for some recycle unconverted MCH, the selectivity is 96.5%.

If the losses in both steps are recovered as the tail gas stream and if this is placed to fuel gas system, this will lower the need to import fuel. This is more significant in the second step. However, the main consequence (cost) of the process losses is that, taken together the 6% (1.5% + 4.5%) toluene process loss and will be required to be made up on each cycle by import of toluene.

In this analysis, toluene is chosen over MCH as the import material because of the extensive and clear trade in the product. Toluene is primarily used as an octane booster for automotive gasoline and there is an extensive trade reported by petroleum reporting agencies. As a consequence of its use its value and traded price are proportional to the traded price of gasoline, and hence varies with the prevailing oil price. For this analysis a value of $800/t is placed on the cost of imported toluene to make up the process losses.

Table 6.7 gives the process economic analysis for both toluene to MCH and MCH to toluene. The quantum of toluene required to carry the

[20] *Ibid.*

373 kt/y of hydrogen is over 5.7 Mt/y. This is used to estimate the required amount of toluene to be purchased to make up for system losses. This is not the toluene inventory, as the system is a cycle with 94% of the toluene being recycled. The daily flow in bbl/d has been used to scale the process plant for both steps of the process.

The individual process plants are assumed to be constructed in 3 years, have a life of 20 years and deliver a 10% discounted cash flow. The capital cost of the first step (toluene to MCH) is estimated at $262M and the second step at $303M.

Following the method used in this work, it is estimated that the cost of producing MCH in step 1 is $101.4/t, which corresponds to a carried hydrogen cost of $483/t, and for the second step of MCH to toluene the hydrogen cost is $800/t.

The significance of the process selectivity is clear from this data. It is assumed that the losses in shipping are zero.

Shipping Toluene and MCH

The estimates for the cost of shipping MCH and back-hauling toluene are developed as previously described and detailed in Table 6.8. Two cases are developed based on the earlier reference DoE methanol shipping study,

Table 6.8: Cost of shipping MCH.

		MCH	**MCH**
Ship capacity	DWT	33,158.21	207,238.8
	GJ		
Capital cost	MM$	51.29	144.54
Operating costs			
Labour	$/day	11073.72	14570.68
	M$/y	3.88	5.10
Fuel	t/day	20	30
	$/t	150	150
	$/day	3000	4500
Port fees/station	$	60,000	80,000

Table 6.8: (*Continued*)

		MCH	MCH
<u>Logistics</u>			
Days/year		350	350
One-way Distance	km	6700	6700
Speed	knots	12	12
	km/h	22.224	22.224
Sailing time	days	12.56	12.56
Turnaround time	h	24	24
One-way trips/year		25.81	25.81
Sailing days/year		324.19	324.19
Port calls/year		25.81	25.81
Days in port/year		25.81	25.81
Return on capital (2 y, const., 15 y life, 10% DCF)	%	15.19	15.19
Capital costs	M$/y	7.79	21.95
<u>Operating costs (Opex)</u>			
Maintenance	%Capex	4%	4%
	M$/y	2.05	5.78
Labour	M$/y	3.88	5.10
Fuel costs	M$/y	0.97	1.46
Port charges	M$/y	1.55	2.06
Insurances	%Opex	15	15
	M$/y	1.27	2.16
Misc (victualing etc.)	%Opex	10	10
	M$/y	0.84	1.44
Total OPEX	M$/y	10.56	18.01
TOTAL COSTS	M$/y	18.35	39.96
Quantity shipped	t/y	427,879.6	2,674,248
Shipping cost (MCH)	$/t	42.88	14.94
Hydrogen equivalent	$/t	700.43	244.03
	$/GJ (LHV)	5.86	2.04

one using a standard 40,000-DWT tanker and the other a 250,000-DWT VLCC-type vessel. The carrying capacities of these vessels have been adjusted for the lower density of MCH. This lower density of MCH relative to toluene sets MCH as determining the capacity of the ships for this duty. The estimates for the shipping cost of MCH are 42.88/t of the 40,000-DWT vessel and $244/t for the VLCC. Since each molecule of MCH contains six atoms of available hydrogen, this corresponds to hydrogen shipping costs of $700/t or $5.86/GJ (LHV) for the smaller vessel and $244/t or $2.04/GJ (LHV) for the VLCC.

Summary

In summary this method of delivering hydrogen results in the following:

- Hydrogen production cost (SMR) including CCS — $1,151/t or $9.63/GJ (LHV)
- Hydrogenation of toluene to MCH (Table 6.7; $1658 – $1151) — $507/t or $4.24/GJ (LHV)
- Shipping (Table 6.8) — $700.4/T or 5.86/GJ (LHV)
- Dehydrogenation of MCH to toluene (Table 6.7) — 800.8/t or 5.86/GJ (LHV)
- TOTAL — $3.159/t or $26.44/GJ (LHV)

When taking into account of the hydrogen production costs, this is considerably lower than the cost of the liquefaction route and avoids the large capital cost of hydrogen liquefaction.

Transport of Hydrogen as Ammonia

Ammonia[21] contains 17.6% by weight hydrogen and this has attracted attention for an alternative method for shipping hydrogen. Ammonia is also widely proposed as an alternative fuel for ships.[22] The ammonia would be synthesised from natural gas, air and water. This is a

[21] Unless otherwise stated, ammonia means anhydrous ammonia.

[22] Gallucci M. (23 February 2021) *IEEE Spectrum*. https://spectrum.ieee.org

well-established manufacturing process (see Chapter 2). It avoids having to produce hydrogen, which would improve the production economics. Alternatively, ammonia may be synthesised using hydrogen produced by electrolysis and nitrogen extracted from the atmosphere. However produced, ammonia would be shipped by tanker to the desired port and then decomposed to hydrogen and nitrogen. The nitrogen component of the ammonia would be discharged to atmosphere and it would not be returned to the point of synthesis. This system would benefit from the known methods of mass production of ammonia for the fertilizer industry and the extensive and large trans-oceanic trade in liquid ammonia. The commercially unproven part of the system is the decomposition of the ammonia into hydrogen and nitrogen, but this should be a facile operation.

The use of ammonia as an intermediary means of producing hydrogen has been researched and developed for many decades. In this work we are concentrating on the mass production and delivery of hydrogen for vehicle refuelling on a large scale and for this role the discussion on the use of ammonia is focussed. Very small-scale use of ammonia as a vehicle fuel may be feasible but, although it has been proposed, it is the author's view that the use of ammonia as a vehicle fuel for the onboard generation of hydrogen will face serious health and safety issues. Recently ammonia has been proposed as a shipping fuel for replacing fuel oil and diesel. This may be feasible but safety considerations would present a major hurdle.

Anhydrous ammonia is highly corrosive and toxic.[23] In small concentrations it has a strong penetrating odour, which heralds its presence but this can become less useful in situations of prolonged exposure. Although ammonia is lighter than air, it is very soluble in water and in humid air forms vapours that are heavier than air. The escape or accidental discharge of ammonia would form a ground hugging vapour cloud. This cloud will for the most part be invisible to the naked eye. A victim entering the cloud would be quickly overcome, collapse and die of asphyxiation. The vapour is flammable. In large-scale producing and using facilities where ammonia discharge could potentially occur, all workers, including office staff, are trained and regularly retrained and tested in using full-set breathing

[23] Material Safely Data Sheets (MSDS) for anhydrous ammonia from the producers and sellers of ammonia give extensive descriptions on the hazards of ammonia use.

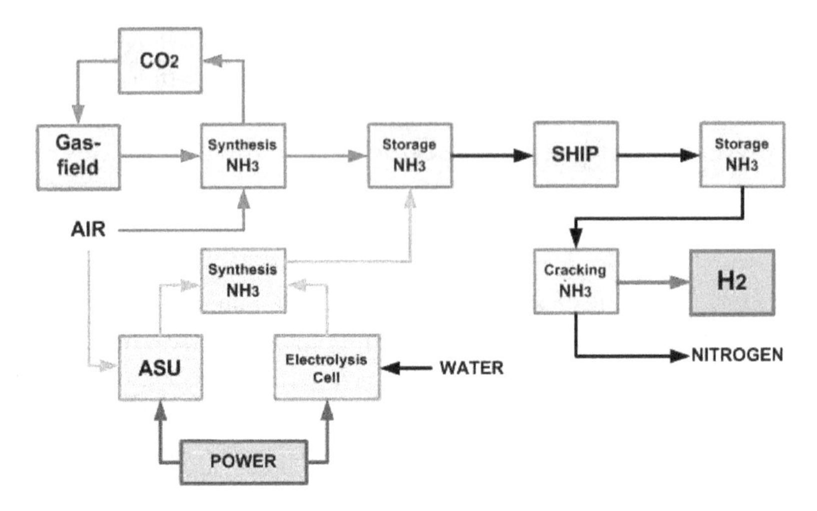

Fig. 6.12: Scenarios for transporting hydrogen as ammonia.[24]

apparatus. For retail vehicle fuelling this would be a requirement at the refuelling site.

Economics of Ammonia Synthesis

The methods of manufacture of ammonia have been described previously (Chapter 2). The block flows of the production by natural gas reforming and water electrolysis are illustrated in Figure 6.12.

In the first method, ammonia is synthesised from a suitable gas field from natural gas by conventional steam methane reforming (SMR) or gasification. Nitrogen enters the system as air in the secondary reforming process.[25] An advantage of this route is that most (about two thirds) of the carbon in the natural gas becomes a concentrated stream of carbon dioxide, which can be compressed and passed back to the gas field for geo-sequestration. If gasification rather than SMR was adopted to manufacture ammonia synthesis gas, all (more than 95%) of the carbon could be geo-sequestrated. Another advantage is that electric power for the process can be produced within the complex from waste heat recovery within

[24] Figure courtesy of Mike Clarke at METTS P/L.

[25] This step could be ammonia synthesis from coal by gasification.

the complex; there may be some excess power that can be exported. After synthesis the ammonia passes to storage for shipment. This scheme is often referred to as 'blue hydrogen'.

The alternative method requires the production of hydrogen by electrolysis. Nitrogen for the synthesis would be provided by an air separation plant. Both of these unit operations require considerable amounts of power and the cost of this impacts on the economics of this route. After conventional synthesis the ammonia passes to storage ready for shipment. This scheme is often referred to as 'green hydrogen'.

Economics of ammonia production by steam methane reforming/gasification

The economics for the production of ammonia are given in Table 6.9. The data is based on a 1990 ammonia facility with a nominal capacity of 850,000 t/y. This is a typical world-scale operation. Some enterprises have double this capacity. Using the methodology is as set out in the Appendix,

Table 6.9: Economics for production of ammonia by Steam Methane Reforming (SMR).

Ammonia		**SMR**
Nominal capacity	kt/y	850
	kt/d	2.50
	PJ/y	19.13
	kt/y H2	150
Base capital Cost	M$	400
Base year		1990
Location factor		1
NF index		2.33
FINAL CAPEX (Capex)	M$	932.52
ROC (3 y const, 20 y life, 10% DCF)	%	14.60
Capital charges	M$/y	136.17
Nominal value	$/t	300.00
Working capital (30 days at $300/t)	M$	22.50

(Continued)

Table 6.9: *(Continued)*

Ammonia		**SMR**
Return of working capital	M$/y	2.25
CAPITAL CHARGES (C)	M$/y	138.42
Operating costs (Opex)		
Labour (1% Capex)	M$/y	9.33
Maintenance (3% Capex)	M$/y	27.98
Insurance (1.5% Capex)	M$/y	13.99
Catalysts and chemicals (1% capex)	M$/y	9.33
NON-FEED OPEX (O)		60.61
TOTAL NON-FEED COSTS	M$/y	199.03
FEEDSTOCK (Gas)	PJ/y	29.62
Thermal efficiency	%	64.58
Feedstock price	$/GJ	2.00
FEEDSTOCK COSTS	M$/y	59.23
PRODUCTION COST CALCULATION		
Gross costs	M$/y	258.27
By-product credits	M$/y	0
Net costs	M$/y	258.27
UNIT AMMONIA PRODUCTION COST	$/t	303.84
HYDROGEN PRODUCTION COST	$/t	1721.77

for a 3-year construction, 20-year lifetime and delivering a 10% discounted cash flow return on the capital cost, which is estimated at $932M (2018). Working capital is assumed to be the value of 30 days inventory at $300/t.

Natural gas feedstock is assumed to be available at $2/GJ and this delivers an ammonia production cost of $304/t. This compares well with typical traded ammonia prices in the range of $400/t to $600/t. The hydrogen equivalent price is $1722/t.

The sensitivity of the ammonia production cost and the hydrogen equivalent cost (ammonia price × 17/3) to the price of natural gas feedstock is illustrated in Figure 6.13.

One of the major thrusts of the hydrogen economy is the elimination of carbon dioxide emissions to atmosphere. As noted above most of the carbon dioxide could be easily compressed and geo-sequestrated.

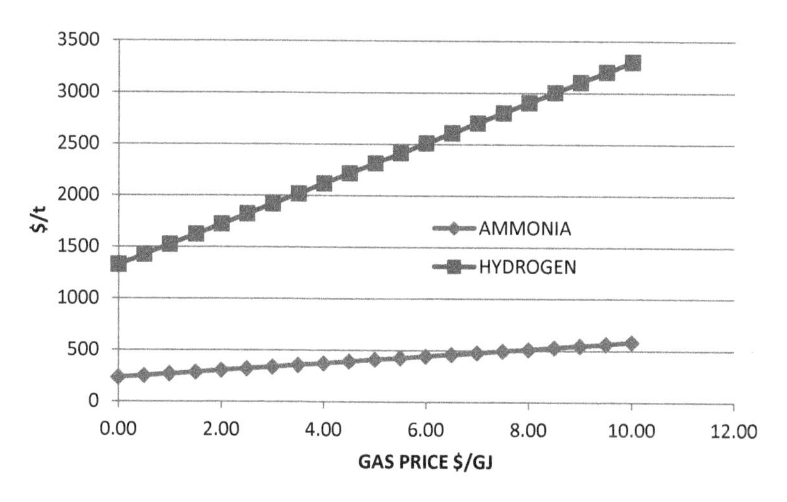

Fig. 6.13: Sensitivity of ammonia and hydrogen production cost to gas price (HHV).

Table 6.10: Impact of carbon dioxide emissions costs on ammonia production cost.

Gas used	PJ/y	29.62
Thermal efficiency	%	64.6
Gas used as fuel	PJ/y	10.49
Carbon dioxide emitted from fuel	$MtCO_2/y$	0.525
Carbon dioxide extracted and emitted	$MtCO_2/y$	0.956
Total emissions	$MtCO_2/y$	1.481
CO_2 cost at $50/t	M$/y	74.04
Non-greenhouse production cost	M$/y	258.27
Greenhouse production cost	M$/y	332.31
Unit production cost (ammonia)	$/GJ	17.38
	$/t	390.95
Hydrogen cost	$/t	2215.4

The other (approximately one third) comes in flue gases in the SMR and could be eliminated if gasification was used. If the latter (flue gas) carbon emission is not captured it could be offset by carbon capture projects elsewhere or the purchase of carbon emission credits. If the cost of carbon dioxide geo-sequestration and carbon emission credits is $50/t (say), the impact on the ammonia production cost is developed in Table 6.10.

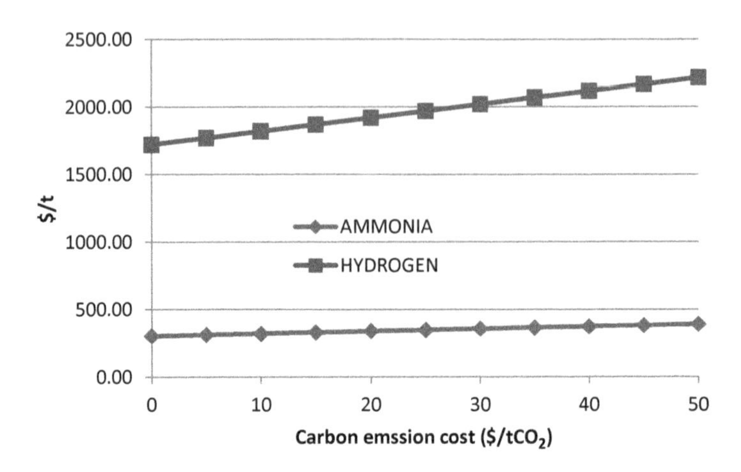

Fig. 6.14: Sensitivity of carbon dioxide emission cost on the production cost of ammonia and hydrogen.

This lifts the production cost of ammonia to $391/t, which is equivalent to a hydrogen cost of $2215/t.

The sensitivity of the carbon dioxide emission cost is illustrated in Figure 6.14.

Economics of ammonia production by water electrolysis

If hydrogen is available from another source such as water electrolysis, then the cost of the process plant for ammonia synthesis will be somewhat cheaper than that for the SMR option. It will not be necessary to have either the primary steam reformer or the secondary (partial oxidation reformer). However, it will be necessary to provide the nitrogen by means of an air separation unit (ASU). The main issue with the ASU is that it would require a significant amount of electric power in addition to the power required for water electrolysis.

Recently, Yara and BASF[26] have announced the opening of a 750-k/y ammonia facility in Freeport, Texas, which uses hydrogen from various petrochemical operations in the region rather than natural gas. This facility was said to cost $600M in 2017. On the assumption that this costs

[26]Yara and BASF open world-scale ammonia plant in Freeport, Texas, *Press Release*, 11 April 2018.

covers all process plant including the ASU and storage as well as the ammonia synthesis unit, this is used as a guide to the cost of producing ammonia with hydrogen produced by electrolysis. The cost of hydrogen by electrolysis has been discussed in Chapter 4 and this data is used as the input for the hydrogen cost. The estimate by this route is given in Table 6.11.

The capital cost has been scaled to 850 kt/y and inflated to 2018. The capital cost is estimated at $693M, which is somewhat lower than the

Table 6.11: Estimated ammonia production cost using hydrogen produced by electrolysis.

Ammonia nominal capacity	kt/y	850
	kt/d	2.50
	PJ/y	19.13
	kt/y H2	150
Construction period	y	3
Required return	%	10.00
Operating period	y	20
Capital recovery	%	14.60
Base capital cost	M$	600
Base year		2017
Scaling factor		1.092
NF index		1.06
FINAL CAPITAL COST (Capex)	M$	696.03
Capital charges	M$/y	101.64
Nominal value	$/t	300.00
Working capital (39 day @ $300/t)	M$	22.50
Return of working capital (10%)	M$/y	2.25
CAPITAL CHARGES (C)	M$/y	103.89
Operating Costs (Opex)		
Labour 1% Capex	M$/y	6.96
Maintenance 3% Capex	M$/y	20.88
Insurance 1.5% Capex	M$/y	10.44

(Continued)

Table 6.11: (*Continued*)

Catalysts and Chemicals 1% Capex	M$/y	6.96
NON-FEED OPEX (O)	M$/y	45.24
TOTAL NON-FEED COSTS	M$/y	149.13
GAS USAGE (hydrogen)	PJ/y	17.93
Efficiency	%	98.00%
Hydrogen price	$/GJ	30.76
HYDROGEN COSTS	M$/y	562.62
PRODUCTION COST CALCULATION		
Gross costs	M$/y	711.74
Byproduct Credits (B)	M$/y	0
Net Costs	M$/y	711.74
UNIT AMMONIA PRODUCTION COST	$/t	837.34
HYDROGEN EQUIVALENT COST	$/t	4744.95

$936M for the SMR route. This route is assumed to have a 98% efficiency on the hydrogen required.

Using the production cost for hydrogen developed above with power available at $50/MWh, the cost of hydrogen feedstock is $30.76/GJ (LHV). This generates an ammonia production cost of $837/t or a hydrogen equivalent cost of $4744/t, far higher than the SMR/gasification route. The sensitivity of the ammonia production cost to the cost of hydrogen is illustrated in Figure 6.15.

It should be noted that the Yara–BASF operation collects and uses hydrogen from several process operations. Presumably this hydrogen source is surplus to requirements at the pertinent facilities in which case we can assume the hydrogen has the value of fuel, which in that location is set by the gas price at Henry Hub. At the time of writing this is about $2/GJ to $3/GJ. With hydrogen at this price the Freeport facility would produce ammonia well below the typical traded price of $500/t.

For hydrogen produced by electrolysis the main parameter of concern is the cost of available power. In this work it is assumed that power costs are in the vicinity of $50/MWh, which seems a common price in many industrialised economies. Many proponents of 'green' renewable energy assert that power is available at $20/MWh. This may be true for

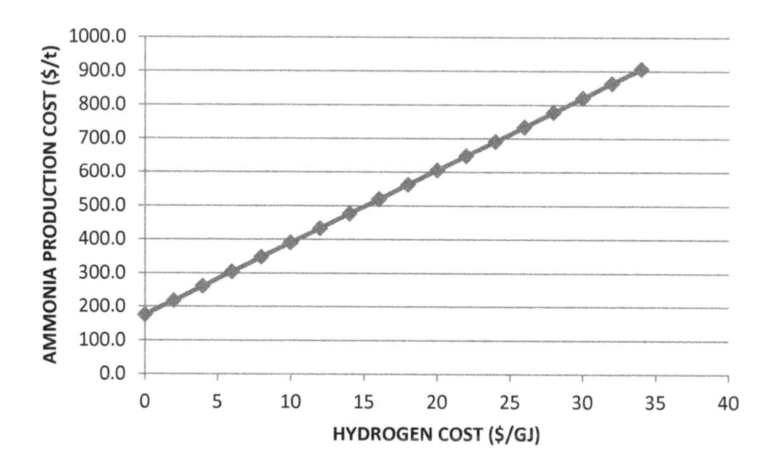

Fig. 6.15: Ammonia production cost sensitivity to the cost of hydrogen.

large-scale hydroelectric schemes, but is unlikely to be true for wind and solar, especially if it requires large-scale battery or an open-cycle gas turbine as backup. Even so at $20/MWh the cost of hydrogen produced by electrolysis is estimated at $16.36/GJ, which generates ammonia at $527/t, which is still higher than the SMR route incorporating carbon capture and storage.

Ammonia Shipping Cost

Anhydrous ammonia is easily compressed into a liquid at ambient temperature. Consequently its shipping is very similar to that for shipping LPG (propane and butane) and the shipping fleet can readily accommodate the shipping of both ammonia and LPG. Some vessels for the ammonia trade can hold 75,000 m^3 of liquid ammonia.

Ammonia fertilizer trade is influenced by the US demand. This is particularly in the corn-belt states of the United States, which is in turn determined by weather conditions in those states. The international shipping trade in ammonia is marked to the price of ammonia from Yuzhny in Ukraine on the Black Sea. This is the result of the historical domination of trans-oceanic ammonia trade by the USSR. The centralised state monetised its vast gas reserves in Russia by pipelining the gas to Ukraine, where it was converted into ammonia and from where it was shipped in

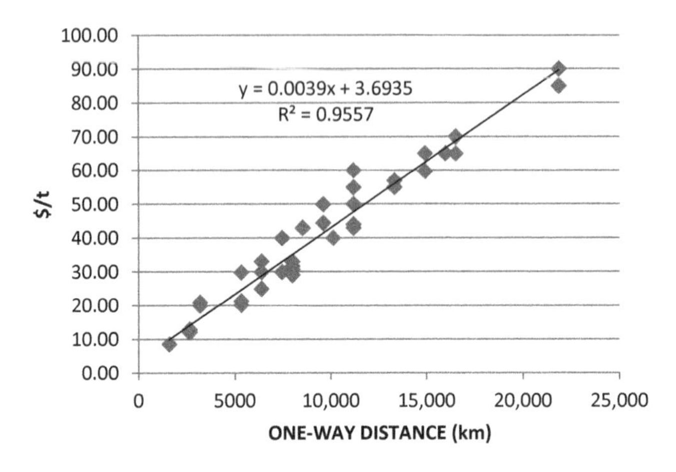

Fig. 6.16: Typical variation for butane shipping costs with one-way distance.

large Soviet merchant marine around the world. Up until the demise of the USSR in 1991, ammonia was more or less available at a constant price (typically $100/t). Since then the market has been more open and complicated by Ukraine–Russian relations, which can affect the availability and price of ammonia out of the Black Sea.

LPG and ammonia shipping costs are widely reported by the appropriate reporting agencies. Costs are proportional to the distance travelled; Figure 6.16 is illustrative of the costs for butane shipping in 2005.[27] In more recent times there has been considerable scatter in the data. Because of the logistics involved, the costs of shipping ammonia is about double that for LPG. The cost of shipping ammonia in September 2018 over a 3500-NM journey (6482 km) was reported to be about $50/t.[28] For this study, the shipping cost of ammonia is taken as $60/t for a 6700-km journey, which is equivalent to hydrogen being carried for $340/t.

The Cracking of Ammonia to Nitrogen and Hydrogen

This is the reverse of the ammonia synthesis reaction. The thermodynamics of the ammonia cracking is illustrated in Figure 6.17.

[27] Data from various sources normalized to a ship capacity of 75,000 m^3.

[28] Argus Ammonia, September 2018.

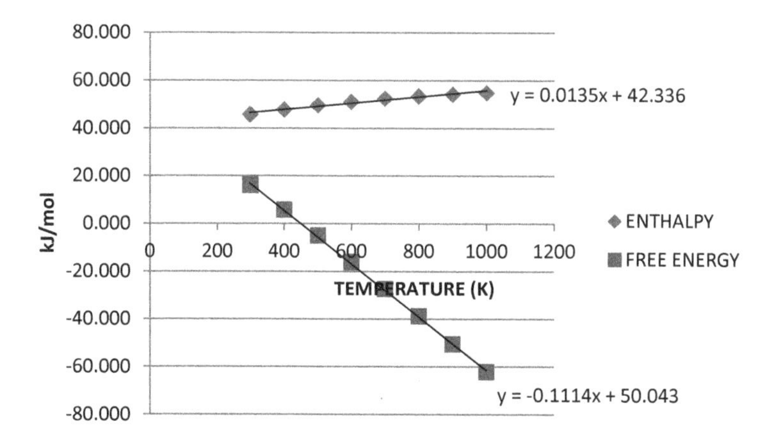

Fig. 6.17: Thermodynamics of ammonia cracking.

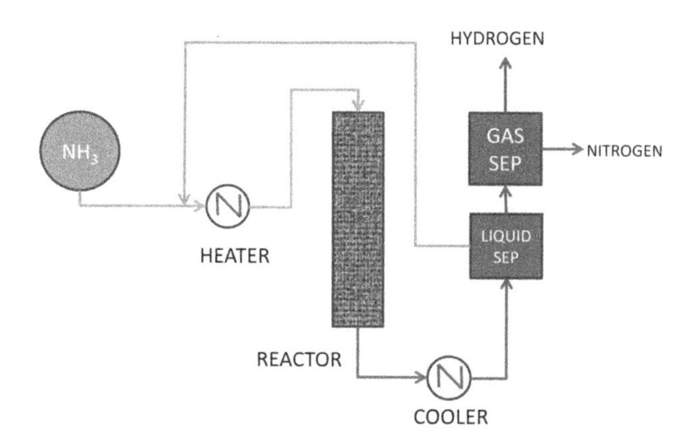

Fig. 6.18: Simplified ammonia cracking operation.

The free energy for the cracking of ammonia into nitrogen and hydrogen indicates that the reaction is feasible at temperatures over about 400 K (130°C). This is a modest operating temperature, however, the main problem is that this will require a considerable amount of energy to be injected into the system with the enthalpy of the cracking being about 50 kJ/mol or about 3 kJ/g (MJ/kg, GJ/t). To this has to be added the heat required to vaporize the ammonia (23.3 kJ/mol) and an allowance for heat losses — say a total of about 80 kJ/mol (4.7 kJ/g, GJ/t) of heat to be supplied.

A simple flow sheet for ammonia cracking is illustrated in Figure 6.18.

In the cracking process, liquid ammonia is taken from storage and heated to the reaction temperature. The duty of the heater is to vaporise the ammonia and provide sufficient heat so as to satisfy the heat required for the reaction. The ammonia passes to the reactor, which is charged with an appropriate catalyst. Ammonia synthesis catalyst is cheap and robust and would be a good candidate. As the reaction proceeds the reactants cool due to the endothermic nature of the process. After cooling, the cracked gases are passed to a separation section. If there is insufficient heat in the system then unconverted ammonia would be present in the exit gases. This will require removal and recycle back to the reactor, possibly by the addition of a separator column and a scrubber unit to further remove unconverted ammonia in the gas stream for recycle. This will be followed by a gas separation section to separate the hydrogen from the nitrogen.

There are several variants proposed for the process. Equinor[29] have proposed a scheme to improve process efficiency by extracting energy from the volumetric expansion of ammonia as it is vaporised. Haldor Topsoe[30] has proposed an auto-thermal-type reactor in which oxygen is introduced into the top of the reactor and part of the ammonia is burnt to provide the reaction heat. This latter concept would have the advantage of not requiring the import of significant quantities of fuel (fuel oil or gas), however, if all the heat was put in by burning ammonia, it would consume 20% of the ammonia feedstock.

There is no economic statistics available for an ammonia cracking operation to produce 373 kt/y hydrogen and so an estimate has been built by an *a priori* method in Table 6.12.

The total cost of the principal items is estimated at $237M to which is added allowances for off-sites, ancillaries and engineering to give a final capital estimate of $332M. The working capital is based on 30 days storage with the ammonia valued at $500/t (typical trade value). The fuel for the operation is taken as fuel oil with a cost of $4.0/GJ. If fuel was to be provided by ammonia the fuel cost would be about $133M/y. The final

[29] Andersen HS, Ingvar Asem K. (21 March 2019) *US Patent Application* 2019/0084831, to Equinor Energy AS.

[30] Speth CH, Wind TL, Dahl PJ. (23 April 2020) *US Patent Application* 2020/0123006, to Haldor Topsoe A/S.

Table 6.12: Estimate for terminal and ammonia cracking unit.

Ammonia storage tanks	M$	$ 137.63
Ammonia cracking unit	M$	$ 100.00
Subtotal	M$	$ 237.63
Off-sites (15%)		$ 35.64
Ancillaries (10%)		$ 23.76
Engineering (15%)		$ 35.64
Capital cost		$ 332.68
Return on capital (2 y const., 20 y life, 10% DCF)		13.84%
	M$/y	$ 46.04
Working capital 30 days @500/t	M$	$ 88.07
Working capital return (10%)	M$/y	$ 8.81
Capital costs (Capex)	M$/y	$ 54.84
Operating cost	%Capex	
Labour	1%	$ 0.55
Maintenance	3.50%	$ 1.92
Insurance	1.50%	$ 0.82
Catalysts and chemicals	1%	$ 0.55
Subtotal	M$/y	$ 3.84
Other variables		
Electricity	M$/y	$ 2.00
Fuel cost (Fuel oil at $4/GJ)	M$/y	$ 49.67
Total production cost	M$/y	$ 110.35
Hydrogen	$/t	$ 295.85
	$/GJ	$ 2.48

result is that the estimate for the hydrogen production cost is $296/t or $2.48/GJ (LHV).

Summary

The following summarises the production cost for delivery of hydrogen by means of ammonia:

- Hydrogen as ammonia by SMR including CCS (Table 6.10) — \$2215.4/t or 18.5/GJ (LHV)
- Shipping as ammonia (Figure 6.15 and discussion) — \$340/t or 2.85/GJ
- Cracking ammonia into hydrogen (Table 6.12) — \$295/t or 2.48/GJ
- TOTAL — \$2850.4/t or 23.85/GJ

This SMR/CCS route using ammonia as carrier compares favourably with the toluene/MCH approach. This is primarily because the hydrogen in natural gas is transferred into ammonia and is not produced as hydrogen in a separate step. However if hydrogen by electrolysis is the choice and the cracking part of the system is driven by ammonia combustion, the toluene/MCH route would be of significantly lower cost.

CHAPTER 7

COMPETITIVE POSITION OF HYDROGEN VERSUS FOSSIL FUELS

The objective of most proposals to develop a hydrogen economy is to eliminate carbon emissions. In order to do this the current suite of fossil fuels needs to be replaced with hydrogen. This covers all aspects of the fuel supply. Of particular interest is the use of hydrogen fuel cells in vehicles to replace the current conventional fuels, particularly gasoline and diesel.

One of the issues for replacement of conventional fuels with hydrogen that will have to be addressed for vehicle fuels is tax. Most, if not all, major jurisdictions receive substantial tax revenue from the present conventional fuels. In the short term, when a hydrogen economy is small and developing, this problem can be largely ignored but in the longer term this loss of revenue will have to be addressed. When and how far a hydrogen economy will be subject to tax is a moot point.

The aim of this chapter is to analyse the competitive position of hydrogen versus conventional fossil fuels and in particular the competitive position of hydrogen fuel cells in vehicles versus conventional transport fuels. To try and analyse this, there is first a discussion of the current state of fuel cells, their operation and the estimation of their efficiency. This can then be used to estimate the cost advantages of hydrogen fuel cells over conventional fuels in a variety of scenarios.

Fuel Cells

The duty of a fuel cell is to take fuel (such as hydrogen) and oxygen (typically air) and by means of electrochemical reactions oxidise the fuel and generate electricity. The fuel combustion produces water and carbon dioxide and for hydrogen fuel only water is produced. Similarly, in electrolysis cells, the chemical reactions are not ideal, and the loss leads to the production of heat in the system. For some large-scale operations this by-product heat can be captured and used elsewhere, much like a combined cycle operation. For vehicles it is necessary to dispose of the heat through a conventional radiator system.

Fuel cells are very similar in construction to an electrolytic cell, effectively operating in reverse; Figure 7.1. Hydrogen (or other fuel) is applied to the anode side of the fuel cell. This generates hydrogen ions (protons) and electrons, typically:

$$\mathbf{H_2 = 2H^+ + 2e^-} \qquad \mathbf{(7.1)}$$

This is a facile process and is typically almost ideal in efficiency. The protons pass through the electrolyte of the cell to the cathode. Air is

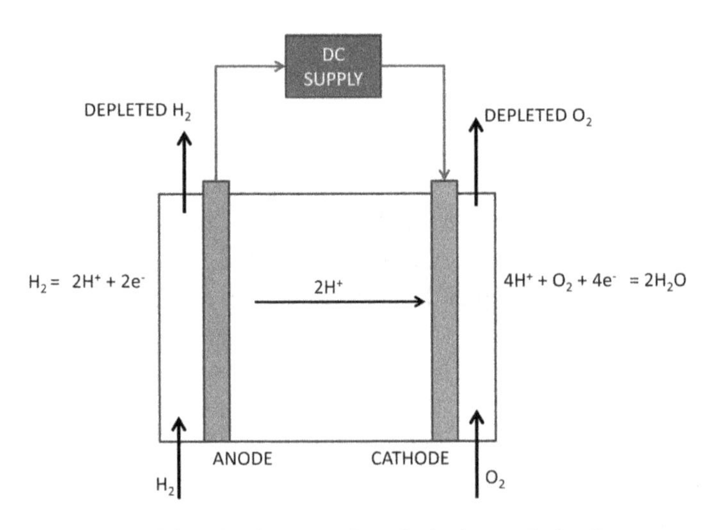

Fig. 7.1: Outline operation of a hydrogen fuel cell.

passed in the cathode side of the cell and the oxygen is reduced to water. At the cathode, the cell reaction is:

$$4H^+ + O_2 + 4e^- = 2H_2O \tag{7.2}$$

By contrast to the anode side, the cathode reaction is complex and it is here that most of the inefficiencies occur. As is the case for electrolysis cells, there is extensive research on the composition of this electrode aiming to improve the efficiencies of the system.

The electrons pass through an external circuit from the anode to the cathode generating the DC voltage to be applied.

As for electrolysis cells, there are several types. These are often divided into those that can operate at relatively low or modest temperatures and those that are more suitable for high temperature operation.

Low to Moderate Temperature Fuel Cells (ambient to 200°C)

Alkaline fuel cells

Operation is typically at ambient temperature. The alkaline fuel cell has been known since 1838. It is durable and has been used in the Apollo space missions. The electrolyte is a solution of potassium hydroxide. The electrodes are composed of porous carbon plates supporting a catalyst, typically platinum.

The anode reaction is:

$$2H_2 + 4HO^- = 4H_2O + 4e^- \tag{7.3}$$

and the cathode reaction is:

$$O_2 + 2H_2O + 4e^- = 4HO^- \tag{7.4}$$

An advantage of these systems in space applications is that the water from the reaction can be extracted and used elsewhere. A disadvantage is that the cell is poisoned by carbon dioxide, which depletes the electrolyte by carbonate formation and has to be excluded from the system. The claimed efficiency using hydrogen as the fuel is over 60%. However, the lifetime of the unit is about 2 years, but recent development may take this to 4 years on continuous operation.[1]

Proton exchange membrane fuel cells

These are commonly abbreviated to PEM or PEFC. They generally operate in the region 60°C–80°C and some are designed for higher temperature operation (140°C–200°C). PEMs have been under development since 1960s and, because of their quick starting and low temperature operation, have largely replaced alkaline fuel cells for vehicle applications. The electrolyte is a fluorocarbon polymer with attached sulphonic acid groups. The electrodes are typically porous graphitised paper, wet proofed with Teflon and backed with platinum black.

The anode reaction is:

$$\mathbf{H_2 = 2H^+ + 2e^-} \qquad \mathbf{(7.5)}$$

and the cathode reaction is:

$$\mathbf{4H^+ + O_2 + 4e^- = 2H_2O} \qquad \mathbf{(7.6)}$$

PEM cells are poisoned by carbon monoxide, which has to be excluded from the system. This is particularly a problem with impure hydrogen or carbon fuels such as methane. They generally have efficiencies in the range of 50%–60%.[2] With hydrogen fuel the efficiency is towards the higher end of the range. The lifetime of the cell is generally in the range 5–10 years.[3]

Phosphoric acid fuel cells (PAFC)

Phosphoric acid fuel cells have been under development since 1960s. The electrolyte is phosphoric acid held in a silicon carbide matrix. The electrodes comprise platinum on porous carbon.

The anode and cathode reactions are the same as those for the PEM fuel cell given above.

They are mainly targeted for stationary applications and very large vehicles and usually have an operating temperature between 140°C and

---

[1] AFC Energy PLC. (2019) *Annual Report.*

[2] *Wikipedia* reports the ideal efficiency is 83%.

[3] Ballard Power Systems. (2019) *Annual Report.*

210°C. The fuel used is typically natural gas and this fuel cell is tolerant to both carbon dioxide and carbon monoxide impurities. They require large loadings of platinum and are therefore expensive. The fuel cell efficiency is low between 37% and 45%, however, the high temperature operation facilitates heat recovery and with this in place the overall efficiency can reach up to 85%. They are claimed to have an operational life of 10 years.[4]

High Temperature Fuel Cells (650°C–1000°C)

The high temperature fuel cells are mainly aimed at using natural gas as the fuel rather than hydrogen. It would be expected that the use of hydrogen would optimise their efficiency.

Solid oxide fuel cells (SOFC)

The electrolyte is a solid ceramic membrane based on zirconia stabilised with yttria. The anode typically comprises a porous nickel coating deposited on zirconia and the cathode is a magnesium doped lanthanum manganate. This fuel cell requires a temperature of at least 500°C to operate and more typically operates in the 800°C–1000°C range.

The anode reactions are for hydrogen fuel:

$$\mathbf{H_2 + O^- = H_2O + 2e^-} \tag{7.7}$$

or for carbon monoxide as fuel

$$\mathbf{CO + O^- = CO_2 + 2e^-} \tag{7.8}$$

for natural gas as fuel

$$\mathbf{CH_4 + 4O^- = 2H_2O + CO_2 + 8e^-} \tag{7.9}$$

and the cathode reaction is:

$$\mathbf{O_2 + 4e^- = 2O^-} \tag{7.10}$$

SOFCs are generally targeted at larger scale electrical utility and distributed power generation applications. They offer fuel flexibility and do not require platinum or other precious metal in the electrodes. They can

[4] Doosan Corporation website: www.doosan.com

suffer from high-temperature corrosion problems and require long start-up times and the number of cold starts can be limited because of stresses imparted onto the ceramic membranes; they are best used on a continuous basis. The efficiency is about 50%. SOFCs degrade slowly with time, typically at a rate of 4% every 1000 h of operation.

Molten carbonate fuel cells (MCFC)

The electrolyte for an MCFC is a molten salt mixture, typically lithium and potassium carbonates, in an inert ceramic matrix. The requirement to have a molten electrolyte defines the operating temperature that is typically in the 600°C–700°C range. The anode is a porous nickel powder alloyed with chromium. The cathode is a porous nickel oxide doped with lithium.

The anode reaction for hydrogen fuel is:

$$\mathbf{H_2 + CO_3^- = H_2O + CO_2 + 2e^-} \quad \mathbf{(7.11)}$$

and for carbon monoxide as fuel

$$\mathbf{CO + CO_3^- = 2CO_2 + 2e^-} \quad \mathbf{(7.12)}$$

The cathode reaction is:

$$\mathbf{O_2 + 2CO_2 + 4e^- = 2CO_3^-} \quad \mathbf{(7.13)}$$

MCFCs are generally targeted for their use in distributed generation operations and are in widespread use for this application. They have an advantage in fuel flexibility, however, they suffer from long start-up times and low power density. Typically, they have sizes of about 2 MW, but there have been proposals for 100-MW installations. Efficiency is said to be about 60% but can reach 85% with heat recovery. MCFCs also have a low decay rate typically <0.5%/1000 h.[5]

Note that in these equations, carbon dioxide is emitted at the anode and requires to be recirculated at the cathode. Because carbon dioxide is

---

[5] Maru HC, Farooque M. (March 2005) *Molten Carbonate Fuel Cell Product Design Improvement*. US DoE, Contract No. DE-FC21-95MC31184 (FuelCell Energy Inc).

produced and then converted into carbonate, MCFCs have recently attracted attention for the capture of carbon dioxide from flue gases.[6]

Fuel Cell Applications

It is not the intention of this work to review in detail fuel cell applications. There have been and continue to be many reviews and conferences of this aspect of the hydrogen economy and any appropriate search will easily locate them.[7] At various points are noted some of the main players and promoters of fuel cells and made reference to their websites. This short review aims to outline the main applications of fuel cells to help address the competitive position of hydrogen versus presently used fossil fuels.

Current Use of Fossil Fuels in Fuel Cells

Putting aside the use of fossil fuels in small electronic devices and the like, we are mainly concerned with uses that consume considerable volumes of hydrogen. These are principally stationary applications and vehicles.

For stationary applications most fuel cells are made specific to the application, can be very large and are principally aimed towards electricity supply and backup. Because the application is generally continuous, with a high demand, the high temperature fuel cells (SOFC, MCFC) are often the ones of choice but this is not universally so. These have the advantage that if the application can utilise the by-product heat, then very high efficiencies can be achieved. Furthermore, these high temperature cells are able to use a wide variety of fuels ranging from natural gas to naphtha as well as hydrogen and are robust to most of the common impurities in the fuel stream. For this use hydrogen would compete directly with fossil fuels such as natural gas. Hydrogen fuel may offer some efficiency benefits over natural gas but considering the overall facility, this benefit will be small.

---

[6] Fuel Cell Energy Inc website: www.fuelcellenergy.com

[7] For instance Nuttall WJ, Bakenne AT. (2019) *Fossil Fuel Hydrogen*. Springer.

Stationary applications are widespread and well demonstrated and most are in the commercial stage of development with many examples. They cover:

- Emergency power backup for private homes, especially where there is poor grid connection or as backup for diesel generators.
- Backup power units for facilities that require power available 24 hours per day, 7 days per week such as hospitals, data centres, police and fire services, grocery and retail. A typical unit will deliver 25 kW for 2 hours.
- To deliver supplementary power to an existing grid. These can be quite large units of 1 MW or more.
- Supply power to residential estates that are not grid connected.
- Provide all power requirements for an existing establishment, such as military bases, hospitals and banks. The fuel cell unit is scaled for the establishment concerned.
- To provide primary power in conjunction with wind generation or solar generation.
- To provide both heat and power to industrial operations and sites.

For vehicles, the fuel cells of choice are generally the smaller AFC, PEM or PAFC types. Larger vehicles tend to favour the larger PAFC type. For these types of fuel cell, a hydrocarbon fuel is usually a lighter boiling fraction of petroleum (such as LPG or naphtha) carried in a conventional vehicle tank. For most efficient use, this is reformed into hydrogen and carbon monoxide and carbon dioxide in an on-board unit, which is like a miniaturised version of steam methane reformer (SMR). There are a variety of designs. For most cases the carbon oxides, because they are deleterious to the operation of the fuel cell, or the hydrogen selectively have to be removed (e.g. by a membrane separator) from the raw reformed gas prior to the fuel cell.

Ignoring carbon emission issues, the approach of using petroleum fractions has been favoured because it uses the existing fuel logistics and infrastructure facilities and can therefore deliver a hydrogen fuel to a fuel cell at low cost. The advantages of the hydrogen fuel cell over conventional fuels for vehicles are:

- Delivers cleaner emissions, 90% lower than internal combustion engines,
- Low noise, although in some instances it can be too low and sound devices are added to warn of the vehicles approach,
- Improved fuel efficiency, more than 100 mpg is claimed,
- Cold start capability with the ability to cold start at –40°C,
- Can be combined easily with batteries to produce a hybrid vehicle,
- Fuel flexibility by hydrogen or other fuels that are processed on board,
- Reduces dependence on imported oil,
- Attracts government support and grants.

The main disadvantages of fuel cell vehicles are:

- The fuelling logistics and infrastructure, especially for hydrogen, are complex and expensive,
- The vehicles are expensive,
- More development is needed to improve efficiencies,
- The vehicles have a limited range compared to conventional fuels, especially diesel.

For a hydrogen economy that seeks to eliminate the present carbon-based fuel system, hydrogen would compete with the conventional transport fuels — gasoline, diesel and fuel oil.

Manufactures of fuel cell vehicles

Fuel cell vehicles have been manufactured as concept demonstration vehicles since 1990s. Most, if not all, of the world's major vehicle manufacturers have produced a concept vehicle.[8] Several manufactures produce production models.

The general layout of a hydrogen fuel cell vehicle is illustrated in Figure 7.2.

There are various designs for the layout of a typical passenger fuel cell vehicle. In the version illustrated in Figure 7.2 the fuel cell is in the

[8]Wikipedia list over 70 concept vehicles by a over 25 manufacturers.

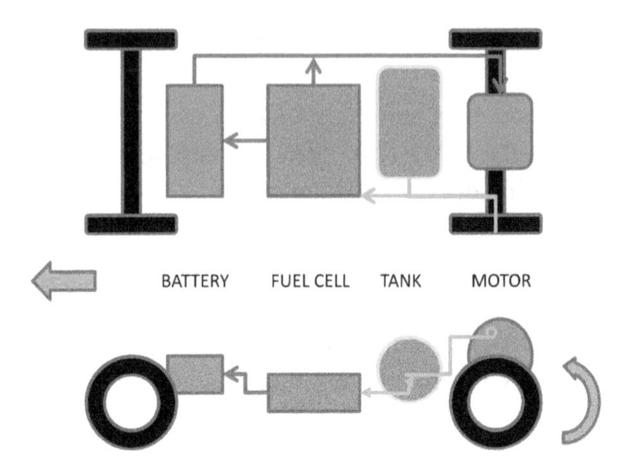

Fig. 7.2: Layout of a hydrogen fuel cell vehicle (after a BMW image).

centre of the chassis and the hydrogen storage tank and electric motor behind. Hydrogen is passed to the fuel cell that generates electricity. This powers the motor and charges a battery that delivers extra power when needed. Not indicated on the figure are electrical inverter units to correct the power supply to the required voltage and the cooling system to remove excess heat from the fuel cell.

As well as passenger vehicles, there have been many larger demonstration vehicles. Buses are common in this regard. The range of buses and the like can be extended by the ability to carry hydrogen fuel in tanks in the vehicle roof. Hydrogen storage is less of a problem with trains and Alstrom[9] have built such trains in the Netherlands.

In 2008, Boeing demonstrated a small electric powered plane powered by a hydrogen fuel cell. Recently (2019) the US DoE announced a $55M investment in aircraft research projects using fuel cells.

Competition for Stationary Applications

In essence the competition between conventional fuels of different types is in terms of cost per energy unit such as $/GJ plus a premium based on

[9]Alstrom. (6 March 2020) *Alstom's hydrogen train Coradia iLint Completes Successful Tests in the Netherlands.* Press Release.

some superior property delivered to the customer or a discount to assist the customer overcoming some pertinent issue. In many instances the specific requirements of the facility dictate the choice of fuels that might be applicable without serious capital expenditure. There are many examples in the power sector.

At one time, coal was the preferred fuel of choice in many jurisdictions for power generation. Coal is found throughout the world (except interestingly the Middle East oil provinces) and power can be generated from a) local source saving import costs. In some parts of the world, fuel oil was preferred because it was easier to store and feed into a large boiler. Some facilities were built to take either coal or fuel oil so that in these markets coal and fuel oil compete head-to-head on price per unit energy delivered. However both coal and fuel oil contain sulphur and in order to limit sulphur emissions (acid rain) natural gas or LNG came into the picture as an alternative. Because gas offered reduced emissions, generators are willing to pay a premium for it so that LNG prices are higher in energy terms than coal or fuel oil. This has led to the construction of large gas-fired generators for base load power using the gas-turbine-combined-cycle (GTCC) technology that offers higher efficiencies than advanced steam boilers but it can only be used with gas or a low boiling (and usually more expensive) liquid fuel such as LPG or naphtha.

In introducing a hydrogen economy, hydrogen will have to find its place in this web of possible alternatives. Of course, the big advantage of hydrogen is that it does not produce carbon emissions so that in broad terms its premium to conventional fuels will be determined by the value of the carbon dioxide abatement to the end-user. In many instances this will be determined by the cost of geo-sequestration of the carbon, any carbon emissions tax or carbon offsets that need to be purchased.

For most large-scale operations, hydrogen would seek to replace coal, fuel oil or heavy diesel. In recent times, the price of thermal coal is typically in the range $50–$100/t[10] for good quality (25–30 MJ/kg) coal. The shipping cost of coal is typically $10/t.

[10] FOB Newcastle, Australia; See OPEC Monthly Report for prices of various fuels including coal.

Table 7.1: Impact of carbon dioxide emission tax on coal energy costs.

CO_2e tax ($/t)	0	20	40	60	80	100
COAL PRICE ($/t)	$/GJ	$/GJ	$/GJ	$/GJ	$/GJ	$/GJ
50	1.67	3.57	5.48	7.39	9.29	11.20
75	2.50	4.41	6.31	8.22	10.13	12.03
100	3.33	5.24	7.15	9.05	10.96	12.87
125	4.17	6.07	7.98	9.89	11.79	13.70

Table 7.1 is used to illustrate the impact of a carbon dioxide emission tax on thermal coal. This gives the relationship between the cost of the coal and carbon dioxide emission tax on the underlying energy cost of the coal feedstock. The coal is assumed to be high quality coal with 30 MJ/kg specific energy and have a carbon content of 78% carbon.

With coal at $50/t and no carbon dioxide emission tax, the cost of the energy in the coal is $1.67/GJ.[11] With a tax of $100 per tonne of carbon dioxide emitted the cost of the energy rises to $11.2/GJ. Now recall the hydrogen cost of production (Chapter 4):

- From gas at $2/GJ with CCS (Blue Hydrogen) — 9.63/GJ
- By electrolysis of water with power at $50/MWh — $30.64/GJ
- By electrolysis of water with power at $25/MWh — $18.68/GJ

Prima facie direct hydrogen substitution of coal without increasing the cost of power generation is viable for blue hydrogen at high coal prices coupled with high carbon emissions costs (coal at >$125/t and carbon emission tax >$60/t). Furthermore, green hydrogen could not substitute for coal without extreme carbon dioxide emission taxes.

Many parts of the developed world have set targets for the elimination of carbon emissions from their jurisdictions by 2050. Few, if any, have developed a road map of how this target is to be achieved, without avoiding many of the predictable consequences — such as increased energy and

[11] Recall that for coal the HHV and LHV are very similar.

power prices, poor food security and deforestation of the natural environment as land is converted to produce renewable fuels. Apart from biofuels, the two other areas touted to achieve the 2050 goal are the mass electrification of industry and the vehicle fleet and the rapid introduction of a hydrogen economy. Another method being proposed is the large-scale introduction of carbon capture and storage from the present operations. This last method is opposed quite vigorously by sections of the public. As noted here substitution of coal-fired generators with hydrogen would require methods that are well beyond those currently under consideration such as carbon emission taxes well over \$100/tCO_2e.

It should also be noted that replacing the current grid generation fleet with hydrogen generators such as SOFC fuel cells and the like is unlikely to produce competitive power prices unless the hydrogen is of the blue type. Because the best fuel cells at present have an efficiency of typically 60%. This is not much better than modern high-efficiency-low-emissions (HELE) coal plants and is similar to the energy efficiencies that can be achieved using integrated-gasification-combined-cycle (IGCC) plants. The basic cost of hydrogen would be too high if the hydrogen were to be produced by electrolysis.

It seems to this observer that the use of hydrogen fuel cells for the large-scale stationary generation is set to be restricted to niche market opportunities rather than in the widespread uptake of the technology for mass power generation.

Competition for Vehicle Applications

For hydrogen-fuelled vehicles, there are two scenarios to be considered, (i) competition against conventional fuels, and (ii) competition against battery technology. In considering the relative competitive positions, because the volume taken up in the vehicle by the fuel is a critical factor, it is important to note the volumetric energy density of the alternatives. This is illustrated in Figure 7.3, which plots the volumetric energy density against the specific energy density for pertinent fuels.

The figure shows the fossil fuel cluster (gasoline, jet-fuel, diesel and fuel oil) have relatively high and broadly similar volumetric and

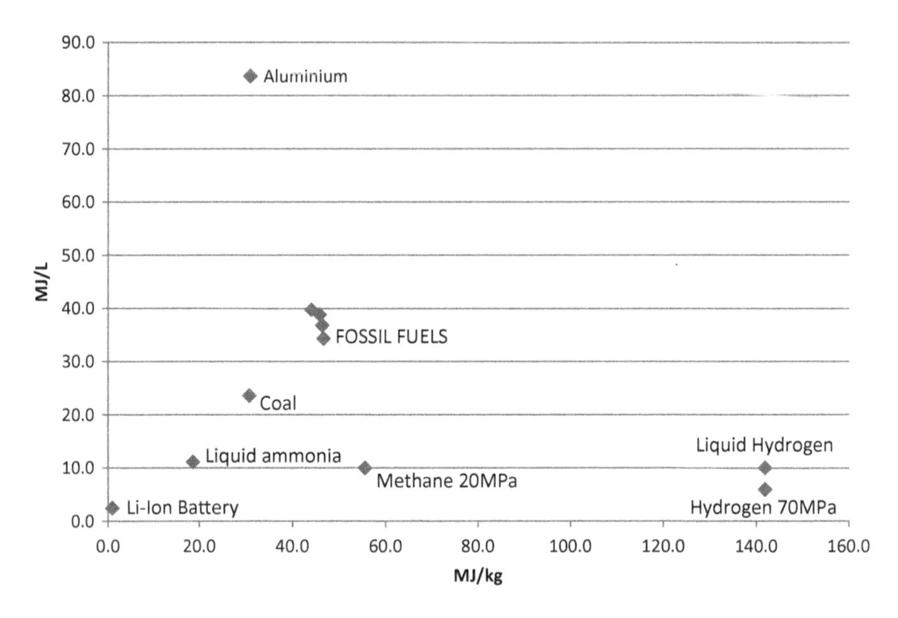

Fig. 7.3: Plot of specific energy density and volumetric energy density for various fuels.

specific energy densities. By contrast, the other fuels of interest (liquid ammonia, compressed methane etc.) have very low volumetric densities. Despite its very high specific energy density, this is also true for hydrogen as either liquid or compressed gas (70 MPa, 700 bar). The consequence is that for substitution of an alternative fuel for a fossil fuel, all other things being equal, for a given volume on-board storage capacity, fossil fuels will deliver higher mileage. A competitive position arises because some of the alternatives, such as hydrogen (and batteries) offer zero carbon emissions and improved engine efficiency to partially offset this problem.

Of note is that today's lithium-ion batteries have very low energy densities and this leads to the mileage challenge for electric vehicles. To offer the same mileage as conventional fuels the challenge for electric vehicles is to dramatically increase the energy storage capacity.

As an aside, the fuel with the highest volumetric energy density is aluminium, which finds its use in rocket engines.

Competitive Position Hydrogen and Fossil Fuels in Vehicles

There are three considerations (i) the price of the fossil fuels, (ii) the relative efficiencies of fuel cells vehicles and conventional engines, and (iii) taxes applying to the fuel sector. These are discussed in reverse order.

Taxes and Local Premiums

Most oil-importing countries levy taxes on conventional fuels. These taxes are for many jurisdictions the source of considerable revenue for governments. In addition they act as a disincentive to excessive fuel use (imports) and incentivise the continued development and use of more efficient vehicles. In recent times there has been a concern with carbon dioxide emissions and this can attract additional fuel taxes from which the noncarbon-emitting fuels are exempt, and as changes to vehicle design rules, which aim to lower the vehicle fleet emissions by continued improvement in efficiencies. The application of emissions taxes to non-fossil fuels can take the form of discounts or waivers from taxes applying to conventional fuels. In many jurisdictions the application of emission taxes and waivers appears somewhat arbitrary.

As to how hydrogen fuel would be treated is moot and would probably be different in each jurisdiction. Since the hydrogen economy is in early stages of development, exemptions from tax and tax waivers might be expected. But if and when hydrogen becomes a significant fraction of the vehicle fleet, then it would be expected that central governments would seek to shore up a dwindling tax on conventional fuels with some form of tax on hydrogen fuels.[12]

The application of taxes and the setting of the basic fuel price to conventional fuels will be illustrated by reference to the Australia and Japan practice for setting the price of gasoline.

Australia has a number of relatively small refineries that account for about half of the market demand for oil products with the rest imported.

[12]The method of raising revenue from an increasing use of electric vehicles which clearly do not pay fuel taxes. is now under active consideration by several jurisdictions.

Dating from the time Australia had a vehicle manufacturing industry, the gasoline (petrol) market is benchmarked relative to 91 octane (91RON) unleaded gasoline. The retail price of the 91RON gasoline is benchmarked relative to 95RON[13] on the Singapore market. Using 95RON as a benchmark builds in a differential favouring the local oil refiners. To this base price is added wharfage (which varies depending on the destination port). To this is added an allowance for distribution and a retail margin. Then excise is added and the whole including excise is subject to a goods and services tax. The sum of this is the retail price for 91RON. Refiners set a premium for the 95RON and 98RON grades. Ignoring the tax components, this sets a local premium relative to the Singapore market of about $10/bbl or about $2/GJ (LHV).

A domestically produced hydrogen fuel would potentially be benchmarked against the wholesale import parity price for 91RON grade, to which will be added a similar allowance for distribution and a retail margin. How much tax (excise and GST) would be levied is debateable.

Japan has a range of major refineries that satisfy the domestic demand. Japan has a major vehicle manufacturing industry producing vehicles requiring 95RON or higher gasoline grades. The Japan retail price for the 95RON grade is benchmarked against the Singapore 95RON grade. There is no premium for the local refiners. To this is added a fuel levy and an import levy. A value added tax (VAT) is then added and an allowance for the retail margin that sets the retail price for 95RON gasoline in Japan.

For hydrogen imported into Japan, on the face of it the price would be benchmarked relative to Singapore 95RON plus shipping costs, the latter is relatively small so that the Singapore price would probably be most important. How much fuel and import levy and VAT hydrogen would attract is again debateable.

Relative Engine Efficiencies

Fuel cell vehicles are generally considered to have a thermal efficiency in the range of 40%–60%. For a hydrogen fuel the efficiency could well be near to 60% mark. This should probably be discounted slightly to allow for transmission losses.

[13] So called 95MOPS.

For conventional internal combustion engines, the efficiency is much lower, typically 25% for many of the older type of gasoline engines. However, in the past 10 years or so, there have been significant improvements with many gasoline engines (Otto cycle) having efficiencies up to 35%, which is typical of a modern diesel engine. However, it should be noted that in some hybrid gasoline/battery vehicles, the engine operates on an Atkinson cycle and efficiencies of 38% are found and claims that an efficiency of over 40% have been demonstrated.

For the purposes of this work it is assumed that hydrogen fuel cell vehicles will operate at 60% efficiency and conventional fuel vehicles will operate at 35% efficiency. This builds in a premium for hydrogen fuel of 60/35 = 1.71 in its price in energy terms relative to the price of conventional fuels.

The Price of Conventional Fossil Fuels

The cost of production of hydrogen from the various sources considered here vary little with the prevailing price of crude oil. The price of conventional fuels are directly related to the prevailing price of crude oil. In brief, there are three major trading areas for oil and oil products, which set the benchmark prices relative a marker crude oil — US Gulf (relative to WTI), North Sea[14] (relative to Brent) and Singapore (relative to Tapis or Brent). The oil market is large, open and transparent with shortages in one area being quickly supplied by others.

Figures 7.4, 7.5 and 7.6 illustrate the high level of correlation between the prevailing price of transport fuels for the Singapore market and the price of Tapis crude oil as the local marker crude oil. The correlations given in the graphs can be used to predict the traded price of transport fuels given a value for the marker crude price, in this case Tapis crude.

To determine the competitive position of hydrogen fuel versus a conventional fuel, the prevailing price of crude oil can be used to predict with a reasonable level of certainty the price of transport fuels prior to the application of local premiums and taxes.

[14] Often referred to as ARA — Amsterdam, Rotterdam and Antwerp.

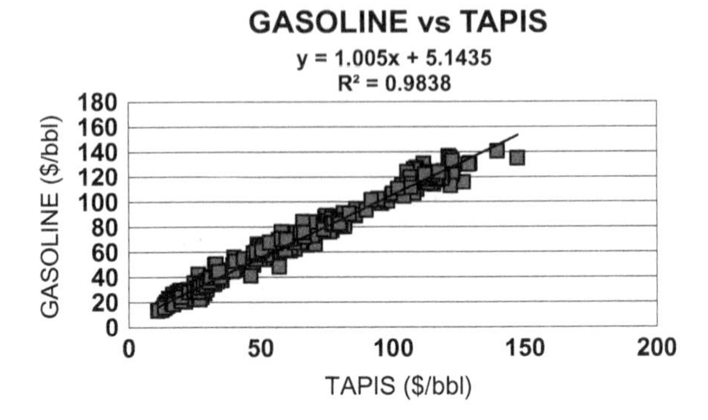

Fig. 7.4: Correlation of gasoline (95RON) with Tapis crude oil on the Singapore market.

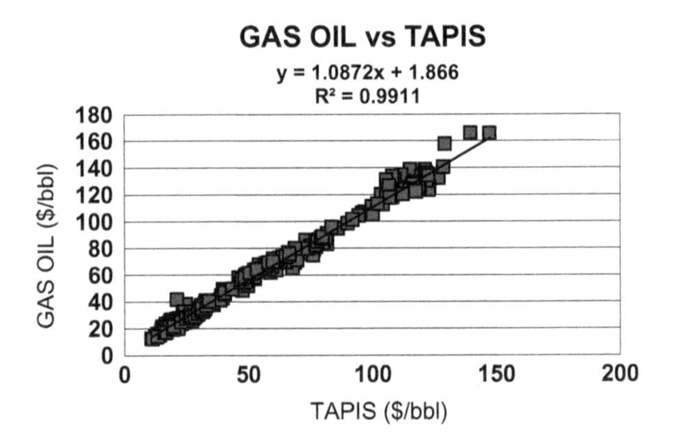

Fig. 7.5: Correlation of gas oil (diesel) with Tapis crude oil on the Singapore market.

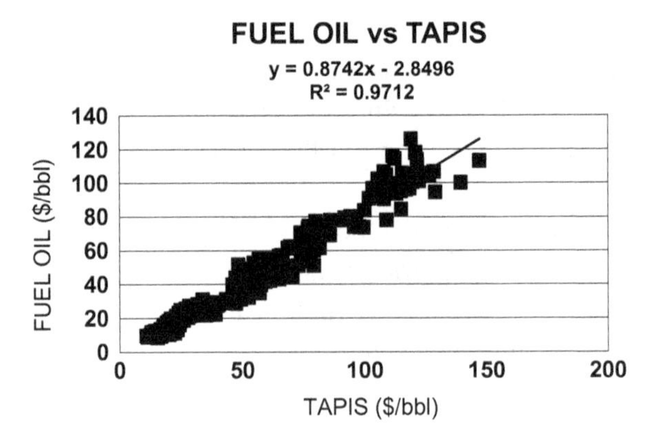

Fig. 7.6: Correlation of fuel oil with tapis crude oil on the Singapore market.

Table 7.2: Estimate for the price of gasoline, gas oil and fuel oil with tapis crude oil at $70/bbl.

Tapis ($/bbl)	70.00		
	Gasoline	Gas oil/diesel	Fuel oil
$/bbl	75.49	77.97	58.30
$/t	646.22	579.32	407.05
$/GJ (LHV)	15.21	13.47	9.47

Table 7.2 illustrates this by estimating the price in energy terms for gasoline, gas oil (diesel) and fuel oil on the Singapore market with Tapis crude oil at $70/bbl.

As noted above, if an adjustment factor of 1.71 is applied to allow for the higher efficiency of a fuel cell vehicle over a conventional internal combustion engine, then the cost of hydrogen production needs to be $26/GJ for gasoline (15.21 × 1.71), $23.1/GJ for diesel (13.37 × 1.71) and $16.2/GJ for fuel oil (9.47 × 1.71). Recall the non-carbon-emitting hydrogen production costs were:

- From gas at $2/GJ with CCS (Blue Hydrogen) — 9.63/GJ
- By electrolysis of water with power at $50/MWh — $30.76/GJ
- By electrolysis of water with power at $25/MWh — $18.68/GJ
- By Biomass feed at $1/GJ — $23.17/GJ.

With oil price at $70/bbl, hydrogen is competitive with gasoline and gas oil/diesel low cost gas incorporating carbon capture and storage, by biomass and by electrolysis with low power prices (<$25/MWh).

Hydrogen is competitive against fuel oil in a motive application, low-speed diesel marine engines, if it is produced from low-cost gas or by electrolysis where power is available at about $20/MWh or less.

Given the errors involved in this analysis, these values indicate hydrogen is quite competitive with conventional fuels when crude oil is priced at $70/bbl and higher. These comparisons take no account of transport logistics and shipping costs. The Chapter 8 will discuss cases that take into account these other costs to obtain a more realistic estimate for the competitive position of hydrogen versus conventional fuels.

CHAPTER 8

CASE STUDIES FOR THE HYDROGEN ECONOMY

This chapter aims to use the data developed in the Chapters 3,4,5 and 6 to evaluate various scenarios in a hydrogen economy and tease out pertinent issues, which will influence a specific location and approach to establishing a hydrogen fuelled solution.

Case Study 1: Hydrogen Production and Use in a Domestic Setting

Most of the work in the previous chapters has focussed on the mass production of hydrogen for powering an economy-wide switch to hydrogen fuel and away from conventional fuels. There is another scenario, which focuses on small-scale and local production of hydrogen for providing domestic power and transportation. In the extreme, this represents a self-contained operation that relies only on water and solar inputs for power.

An outline of this scenario is given in Figure 8.1.[1] This outline is developed for the author's location in Melbourne, Australia.

The inputs to the concept are solar radiation, which is converted into electricity by a solar array, and water. There is a grid connection for back-up or export of excess power as may be necessary. The solar array feeds

[1] Diagram courtesy of Mike Clarke, METTS Pty. Ltd.

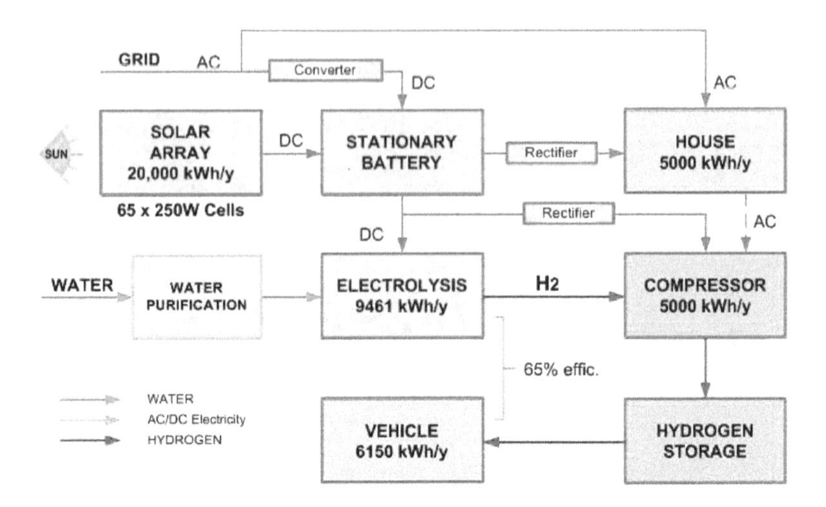

Fig. 8.1: Domestic hydrogen economy.

electricity to a battery system, which acts to hold excess power generated and act as a modulator for the power supply. The battery system is also connected to the grid. The system to this point is becoming increasingly popular for providing the electricity needs of a household.

In a domestic hydrogen economy, the solar battery system also supplies power for an electrolysis unit, which generates hydrogen. This electrolysis unit is supplied by potable water but will almost certainly require pre-treated to remove chloride and similar anions, which would otherwise disturb the electrolysis. Depending on the water quality, it may be necessary to remove cations, if these would interfere with the hydrogen production.

The primary purpose of the hydrogen is for vehicle transport. However, an alternative approach would be to operate a larger electrolysis system and use hydrogen to power a fuel cell for domestic electricity production. This option is not shown in the figure.

For a vehicle operation, compressed hydrogen is required. Although 70 MPa (700 bar) would be preferred, for a domestic operation a lower pressure of 35 MPa may suffice. The figure indicates a compressor and there are commercially available units operating with diaphragm compressors for this duty. In future, this may be a membrane compressor.

Whatever compressor type is chosen, it is unlikely to achieve the high efficiencies of large-scale reciprocating piston compressors, so that whatever is chosen the power demand for the compressor would be quite high. After the compressor, the hydrogen would be stored in cylinders and be available for charging the on-board fuel storage system in the vehicle.

The estimates on the figure are based on a vehicle requiring 3.55 kg of hydrogen per week, which is theoretically about 19,000 km/y (12,000 miles/year). This translates to 6,150 kWh/y which means that if the electrolysis cell is operating at 65% efficiency (alkaline cell), the electrolysis will consume 9461 kWh/y. Allowing 5000 kWh/y for the compression and a similar amount for the electricity demand for the household (typical for Victoria in Australia), the solar array would need in theory to provide 20,000 kWh/y. At the time of writing this could be achieved by 65 × 250 W solar cells. This is very large for a domestic solar array, but may be feasible for a remote farm site and the like.

One of the basic assumptions is that the solar irradiance is constant. This is not so as there are clearly seasonal variations. The variation of solar panel output with season is generally available from the major manufacturers and purveyors of solar cell systems from which the data for the author's location was adapted and used to generate Figure 8.2.

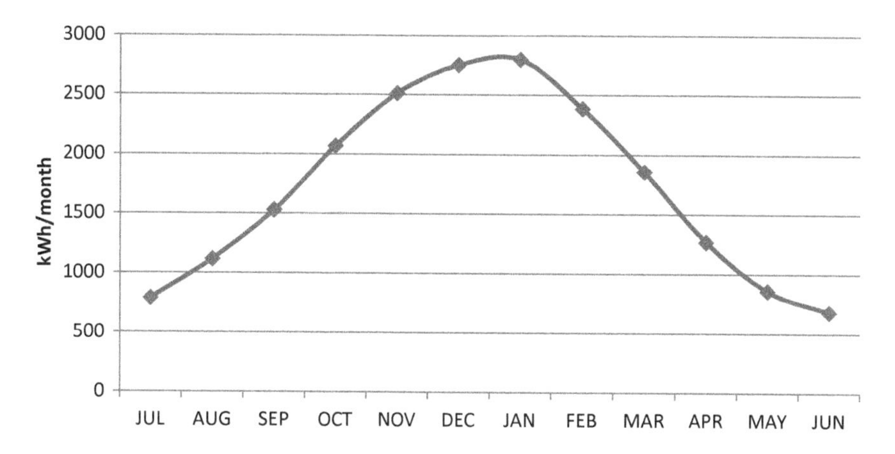

Fig. 8.2: Monthly output for a system capable of generating 20,000 kWh/y; Melbourne, Australia.

Inspection of Figure 8.2 clearly demonstrates a major problem is that although the system will deliver 20,000 kWh/y or an average of 1667 kWh/month, over the year there are large swings in output from under 1000 kWh/month (May to August) to over 2,500 kWh/month (November to February). For the domestic hydrogen scenario there are three possible solutions (i) large increase in the number of solar panels to deliver the required power at periods of minimum solar irradiance and appropriate scaling of equipment to cope with the times of high solar irradiance, (ii) large increase in the storage volume for hydrogen and/or the battery, and (iii) grid connection to import or export power as required. The first two solutions would deliver off-grid power and hydrogen but only at higher cost. The third option requires the presence of a grid connection to provide power at times of minimum solar output and possibly take excess power generated in the summer months.

Taking into consideration these ideas and points, it is highly likely that this domestic hydrogen economy solution will not be a low-cost option for providing domestic power and transport.

Case Study 2: Hydrogen from Large-Scale Solar Power

The concept of a green hydrogen economy is to produce hydrogen by electrolysis, which is powered by some sort of renewable generation. This has been practiced for many years in those parts of the world which have available low-cost power from hydropower generators such as Norway. More recently, there have been proposals to use other types of renewables such as wind generators or solar power. In places with high levels of solar irradiance, such as central Australia or the Middle East, there have been proposals to generate hydrogen by coupling the electrolysis with large-scale solar generators. Some of these schemes use different forms of zero carbon emission power, e.g. wind generators being backed up with hydro-power to deliver constant power for the electrolysers.

A concept for the use of large-scale solar is considered in this case study. The idea is to generate electricity by means of a solar array in a remote area, which has a high level of solar irradiance and where land is relatively cheap to hold the large area required for the solar arrays, possibly many hundreds of hectares holding several million solar panels. The concept envisaged produces sufficient hydrogen for a central facility servicing 40,000 vehicles, i.e. about 373 kt/y of hydrogen production. There are several possibilities some of which are considered here.

- Electrolysis and compression at the site and sending the hydrogen by pipeline to a city filling facility,
- As above but to utilize an existing natural gas pipeline,
- As above but to transmit the produced electricity to a city location by high-tension cable and then produce hydrogen in the city environs.

The principal problem to be addressed is the production cost of solar power on large scale in a remote site. Examination from public domain information of recently (2017–2020) proposed developments in eastern Australia (South Australia, New South Wales and Queensland) have produced the statistics given in Table 8.1.

The average capital cost of the facilities was AUS$M1.29/MW. Some of the proposals also had battery back-up, but this appeared to be mainly to regulate the output rather than to provide power on an extended basis. Including batteries virtually doubles the capital cost. The average efficiency of the schemes was 24.2%. This was estimated from the declared annual MWh produced or inferred from the number of homes to be theoretically supplied — the average domestic consumption of electricity in different states in Australia is reported by the government regulator — compared to the declared capacity in MW. The efficiency value (24.2%)

Table 8.1: Some statistics for large-scale solar in Australia.

No. of proposals considered	18
Lowest capacity in MW (AC)	133
Highest capacity in MW	1500
Lowest number of solar panels	500,000
Highest number of solar panels	1,000,000
Average Capital cost AUS$M/MW	1.29
Average capital cost with battery AUS$M/MW	2.10
Average efficiency	24.2%

seems high for a solar power operation, but is accepted for the purpose of this analysis.

These values have been used to scale a solar operation to power an electrolysis plant capable of producing sufficient hydrogen for a central fuelling facility — 373 kt/y. The statistics for this generator is developed in Table 8.2 for a system using proton exchange membranes (PEM) cells operating at 58% efficiency.

The power required to generate the 373 kt/y hydrogen is 2629 MW. If this has to be supplied by the solar array, then at the 24.2% efficiency an array of 10,871 MW of installed capacity is required. The capital cost of this array has been developed from the data in Table 8.2 assuming that the cost can be reduced by 10% by not requiring the inverters to convert the produced DC (direct current) in AC (alternating current). This reduces the capital cost to AUS$M1.16/MW or with AUD/USD = 0.7, US$M0.814/MW.

Table 8.2: Estimate for power generation cost for large-scale solar array (US$).

Power required	MW	2628.72
Efficiency	%	24.2%
Required installed capacity	MW	10,871.776
CAPEX	M$US,	8,851.2067
ROC (2 Y 20 y life, 10%Dcf)	%	13.57%
ROC (2 Y 20 y life, 10%Dcf)	M$/y	1200.80
Working Cap.	M$/y	0
OPEX		
Labour (20)	M$/y	2.00
Maintenance (1% capital cost)	M$/y	88.51
Chemicals	M$/y	2.00
Insurance (1.5% capital cost)	M$/y	132.77
Total OPEX	M$/y	225.28
Total Costs	M$/y	1426.09
Output	MWh	2.15E+07
Unit Power cost	$/MWh	66.48

Using the statistics from Table 8.1, this is an enormous array and similar in size to that proposed for the Sun Cable project to supply power to Singapore.[2] It is assumed that this can be built in 2 years, has a 20-year life and delivers a 10% discounted cash flow on the investment; the return on capital is 13.57% or M$1,200/y. Once up and running solar arrays appear to have low operating and maintenance costs.

The total costs are M$ 1,426/y which works out at $66.5/MWh. This value is close to that reported in the literature for large-scale solar costs.[3]

Using the method previously used for estimating the hydrogen production cost by electrolysis the estimated on-site hydrogen production cost sensitivity to power cost is shown in Figure 8.3.

At $66.5/MWh, the estimated hydrogen production cost including the capital component is $4,625/t or $38.7/GJ (lower heating value [LHV]). Removing the capital component, the cash cost of hydrogen production is estimated at $3906/t or $32.7/GJ (LHV). This re-emphasises the importance of minimising the power price for solar power generation.

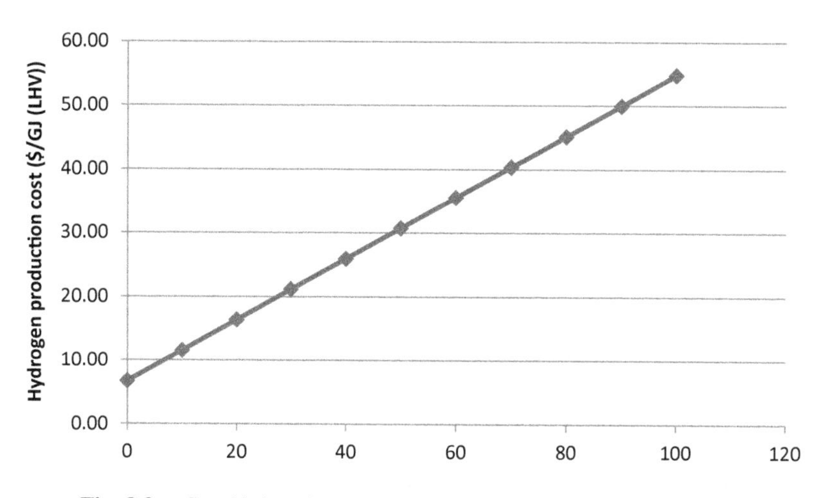

Fig. 8.3: Sensitivity of hydrogen production cost to electricity price.

[2] Sun Cable envisages a 10 GW array covering 12,000 hectares near Tennant Creek in the Northern Territory of Australia. This is coupled with 30 GWh of battery storage and is aimed at supplying 20% of Singapore's electricity needs: www.suncable.sg

[3] P. Zubrinich, Solar costs have fallen 82% since 2010, *PV Magazine* June 4, 2020

Table 8.3: Compression cost estimates for pipeline hydrogen.

	Input Bar	Output Bar	Power MW	Capex M$	ROC M$/y	$/GJ
Electrolysis to distribution	1	200	352	2547.0	345.5	7.73
To High-Pressure Distribution	200	800	37.8	273.5	37.1	0.83
End of Pipeline to High-Pressure Fill Station	100	800	65.4	473.2	64.13	1.44
To Low-Pressure Distribution	200	400	16.4	118.7	16.0	0.36
End of Pipeline to Low-Pressure Fill Station	100	400	37.8	273.5	37.1	0.83

If the hydrogen is to be delivered by pipeline it will require compression. For a central facility (373 kt/y hydrogen), the compression costs are estimated for various options of inlet and outlet pressures in Table 8.3. The compression costs have been estimated using cost of $7235/kW and the return on capital assuming a 2-year construction, a 20-year life and delivering a 10% Dcf.

Option 1: Hydrogen Delivery by Pipeline

The first option to be considered is the production of hydrogen by solar power at a remote site, 1600 km from the central distribution facility, this is followed by compression and transport by pipeline to the central distribution facility. The outline statistics for the pipeline option are given in Table 8.4.

With a cost of M$1/km, which is reasonable for a remote site in a single jurisdiction and passing through open country, the total cost of the line is M$1,600. For a 2-year construction period, a 20-year lifetime and delivering a 10%Dcf the return on capital is M$217.07/y. The pipeline is assumed to not require recompression and is available at 20 MPa (200 bar) at the remote site and is delivered to the distribution point at 10 MPa (100 bar). The operating costs are estimated at 2% of the capital cost. The pipeline portion of this option delivers hydrogen at a cost of $5.57/GJ (lower heating value [LHV]).

Table 8.4: Pipeline cost for transport of 373 kt/y hydrogen.

Distance	km	1600
Cost	M$/km	1
Total cost	M$	1600
Return on capital (ROC) (2 y; 20 y, 10%)	% Capex	13.57%
Return on capital	M$/y	217.07
OPEX (2% Capex)	M$/y	32.00
Total costs		249.07
Hydrogen	PJ (LHV)	44.68
	$/GJ (LHV)	5.57

The total estimated cost of the delivery of hydrogen to the central facility by this option can now be estimated by summing the cost estimates for the various parts:

- Cost of production of hydrogen by solar power at the remote site — $38.7/GJ
- Compression to 20 MPa (200 bar) — $7.73/GJ
- Pipeline transport cost — $5.57/GJ
- Compression from 10 MPa to 80 MPa (800 bar) — $1.44/GJ
- TOTAL ESTIMATE HYDROGEN DELIVERY — $53.44/GJ

Option 2: Hydrogen Delivered by Existing Pipeline

For this option it is assumed that a natural gas pipeline exist and can be utilised for the carriage of hydrogen as a pure component or as a mixture with natural gas. For this option, an allowance is made for hydrogen losses in the distribution system and for extraction of hydrogen at the delivery point.

For this estimate, it is assumed that the cost of hydrogen carriage is based on the carriage cost of natural gas on a volume carried basis. If the carriage cost for natural gas over a 1600 km line is $1.5/GJ or $1,073/ Mm^3, then the hydrogen tariff is $0.12/GJ. Comparison with the cost for the transport over a new pipeline (Option 1) clearly illustrates the attractive idea of using established pipelines for hydrogen carriage — with the caveat that all other things are equal. The cost for this option is estimated as:

- Cost of production at the remote site $38.7/GJ
- Compression to 20 MPa (200 bar) $7.73/GJ
- Pipeline transport cost $0.12/GJ
- Compression from 10 MPa to 80 MPa (800 bar) $1.44/GJ
- Loss allowance $1.00/GJ
- TOTAL ESTIMATE $48.99/GJ

Although the total cost is dominated by the hydrogen production cost, using an existing pipeline could offer significant savings against a dedicated pipeline for hydrogen transport.

Option 3: Delivery of Power from Remote Solar Farm by Transmission Line to Central Hydrogen Production Facility

This option envisages a remote solar generation facility producing power which is transmitted to the central distribution facility by high voltage (400 V DC) electric cable and the hydrogen is then generated at the central facility. This option relies on the lower cost of constructing a transmission cable rather than a new dedicated hydrogen pipeline. Assuming cost associated with producing the high voltage is subsumed in the transmission line costs, the salient statistics are given in Table 8.5.

The cost of the transmission line is assumed to be M$0.5/km.[4] It is assumed that there are no transmission losses and the voltage changes at the end are incorporated into the central facility costs. The cost of the power transmission is $4.16/MWh. If this cost is applied to the production cost of hydrogen, it is estimated that it will result in an increase in the cost

---

[4] See 'ElectraNet Transmission Line Cost Review' Jacobs for ElectraNet Pty. Ltd. RO139100-EE-REP-0002/D, February 2019.

Table 8.5: Cost of power transmission from a remote site to a central hydrogen production facility.

Distance	km	1600
Cost	M$/km	0.5
Total cost	M$	800
ROC (2 y; 20 y, 10%)	%	13.57%
Return on capital	M$/y	108.53
Power	MW	2980.72
	MWh	26,111,068
Power cost	$/MWh	4.16

of production of hydrogen to $4863 or $40.71/GJ. The total estimated for this option is

- Cost of production using power from the remote site $40.41/GJ
- Compression to 20 MPa (200 bar) $7.73/GJ
- Compression from 20 MPa to 80 MPa (800 bar) $0.83/GJ
- TOTAL ESTIMATE $48.97/GJ

This is lower that the estimate using a dedicated hydrogen pipeline and similar to the cost for transporting hydrogen by an existing natural gas pipeline.

The relative merit of this option may be further enhanced by consideration of the capital costs. The capital and operating costs of producing hydrogen at a remote site would be expected to be significantly higher than manufacture near an urban centre.

Comparison to Conventional Fuels

The previous chapter discussed the cost differentials for fuels by comparing fuel cell vehicles with conventional vehicles. This included a discussion of the impact of the prevailing crude oil price on these differentials and the benefits arising to hydrogen fuels as a consequence of the higher efficiencies of fuel cells compared to those of internal combustion engines. At $70/bbl for a local market crude oil such as Tapis crude oil, the cost of gasoline and diesel is $15.2/GJ (LHV) and $13.5/GJ (LHV) respectively. If a local premium is applied for an Australian situation ($2/GJ for gasoline and $1/GJ for diesel) and making an allowance for the higher efficiencies of fuel cell vehicles, a competitive target price for hydrogen is $29/GJ for gasoline and $24.9/GJ for diesel. This is on an untaxed basis.

The costs developed here for remote hydrogen are well above these levels which means that, with oil at $70/bbl, hydrogen fuel delivered by these means are uncompetitive against the conventional transport fuels. To make hydrogen competitive would require a large increase in tax on the gasoline, tax waivers against excise, high carbon taxes or subsidies for hydrogen production or a combination of all of these.

Case Study 3: Remotely Produced Hydrogen Shipped to North East Asia

At the time of writing, Japan is heavily promoting the use of hydrogen for its economy. In this case study, the production of hydrogen in the North West of Australia and its trans-oceanic shipment to Japan is examined. This case is benefitted by the presence of large established liquefied natural gas (LNG) facilities in the North West of Australia and the presence of supporting infrastructure, which in turn rely on the relatively cheap and plentiful reserves of natural gas. Furthermore, any carbon dioxide produced from options for this case could be, in theory, geo-sequestrated in depleted oil and gas reservoirs.[5]

Despite recent proposals to the contrary,[6] it can be deduced from the data on the above case study that the generation of hydrogen from solar power, is unlikely to be the best option. For this to present a viable option on the statistics reported here, there will have to be a major reduction in solar power generation costs and a significant improvement in the efficiency of the electrolysis. As a consequence, this case study concentrates on the production of hydrogen from natural gas. This is followed by

[5] Note the large Gorgon carbon dioxide geo-sequestration project is in this region.

[6] W. Fowler. (23 October 2020) Renewable energy hub to get major project status. *Australian Financial Review.*

examining the alternative means of shipping hydrogen to Japan. Also considered is the potential to ship the hydrogen as LNG and convert the LNG to hydrogen in Japan.

The use of natural gas as a feed for hydrogen and geo-sequestration of co-produced carbon dioxide is often referred to a blue hydrogen option. This is similar to blue hydrogen projects proposed for the Rotterdam area of The Netherlands.[7]

For this estimate, the approach is as explained in the foregoing chapters. Natural gas is assumed to be available at $2/GJ and to produce 373 kt/y of hydrogen, two large-scale methane steam reformer complexes would be required. By-product electricity is assumed to be valued at $50/MWh.

The construction of facilities in a remote area would also imply larger capital and operating costs. A 50% loading is added to cover the increase in cost due to the location. The production cost from a steam reformer option for producing hydrogen rises from the base case developed in Chapter 4 of $584/t to $735/t or $6.12/GJ (LHV).

Carbon dioxide (4.24 Mt/y) would be emitted and required to be geo-sequestrated. This would be similar in size to the existing Gorgon project in this region. If the cost of this emission by geo-sequestration or the purchase of carbon credits is $50/t, this will increase the cost of production to $10.55/GJ (LHV).

Option 1: Shipping Hydrogen as Liquid and Regasification

As discussed in the previous chapters, liquefying hydrogen on a mass scale in an analogous manner to LNG involves very high capital and operating costs. Using several modules of smaller-sized plant may lower the cost. Using the cost liquefaction estimated by US DoE for this type of system, the cost of liquefaction is developed in Table 8.6.

Shipping and regasification costs are as developed in the earlier chapters. In summary this approach delivers hydrogen at a cost of:

---

[7] Anon. (2019) H-vision in Rotterdam targets blue hydrogen from natural gas. *Fuel Cells Bulletin* **3:** 9

- Natural gas to hydrogen with carbon capture and storage (CCS) $10.55/GJ
- Liquefaction $24.78/GJ
- Shipping (Chapter 5) $2.91/GJ
- Regasification (Chapter 6) $6.45/GJ
- TOTAL DELIVERY COST $44.69/GJ (LHV)

This cost is lower than the cost developed for a remote hydrogen facility and distribution using solar power within Australia. This is a consequence of the lower mass transport cost (shipping vs. pipelines) and the absence of this case of compression costs. If compression cost is added to

Table 8.6: Hydrogen liquefaction costs after US DoE for remote NW Australia.

Capacity	t/d	1100
Capex (see Chapter 6)	M$	2755
Location Factor		1.5
CAPEX (at location)	M$	4544.1
ROC (2 y, 20 life, 10% Dcf)		13.84%
Return on Capital	M$/y	628.81
OPEX	% Capex	
Labour	1%	45.44
Maintenance	3.00%	136.33
Catalysts and Chemicals	1%	45.44
Insurance etc.	1.50%	68.16
Total		295.37
Power	kWh/kg	10
	MWh/y	3,739,400
	$/MWh	50
	M$/y	186.97
TOTAL COSTS	M$/y	1111.16
	$/t	2971.49
	$/GJ	24.78

deliver the regasified hydrogen at pressure, the estimated cost would be over $45/GJ.

Option 2: Shipping Hydrogen as LNG

This option considers using the benefits of the existing infrastructure to ship hydrogen as LNG to Japan and then converting the LNG into hydrogen there. Carbon capture and storage would also be performed in North East Asia.

The landed price for LNG traded into Northeast Asia is well known and widely reported. Generally speaking, the LNG landed price is dependent on the prevailing crude oil price. As a consequence of the price volatility in the oil market, the landed price of LNG varies from about $6/GJ (higher heating value [HHV]) to sometimes over $14/GJ (HHV). For this option, it will be assumed the landed price of LNG is $10/GJ. To this is added the regasification cost of $1.12/GJ (HHV) for a land-based regasification option. This could be reduced to $0.68/GJ (HHV) for a floating storage and regasification unit (FSRU) (see Chapter 6).

Northeast Asia has a well-established and efficient hydrocarbon processing industry. If we assume that the cost of production of hydrogen is similar to the US Gulf, then the cost of hydrogen production would be $2,651/t or 22.10/GJ (LHV) with a gas input cost of $11.12/GJ (HHV).

For delivery of hydrogen product which contains no embedded carbon dioxide as a consequence of its production, some cognizance should be taken of carbon dioxide emissions in producing the LNG. Many natural gases contain very low amounts of carbon dioxide. In some gas fields, the carbon dioxide content is high (e.g. the Gorgon project on the North West Shelf of Australia) and the carbon dioxide in the gas is stripped from the gas and reinjected in carbon capture and storage operations. However, a significant level of emissions arise from the energy used in the LNG facility itself to drive the refrigeration plant. This energy uses, is in most cases, some of the input gas to drive gas-turbine generators and often this is an open-cycle gas turbine. These produce a flue gas with is very low in carbon dioxide and consequently is difficult capture and geo-sequestrate. For a zero emissions operation, this will have to be offset by carbon emission credits or the power provided from a non-fossil fuel source.

In order to estimate the emission from LNG production, it is assumed the whole facility efficiency is 85%.[8] This allowance should also be sufficient to cover other emissions such as those arising from emissions in shipping the LNG. Making this allowance increases the amount of gas required to produce the hydrogen. This makes the estimated hydrogen production cost $3,066/t or $25.56/GJ (LHV).

To this should be added the cost of geo-sequestration of the carbon dioxide produced during the production of the hydrogen. At $50/t of carbon dioxide, this will increase the cost of hydrogen production to $3,733.1/t or $31.12/GJ (LHV).

In Northeast Asia, the preferred carbon dioxide disposal option may be in deep sea trenches, which are found in the region. The cost of geo-sequestration in these sites may be somewhat higher than the more conventional depleted oil and gas reservoirs. If this disposal cost is $100/t of carbon dioxide, the hydrogen production cost will rise to $4,400/t or $36.68/GJ (LHV).

Option 3: Transport as Ammonia to Japan

For this option, the natural gas is converted to ammonia at the site and then shipped to Northeast Asia using the available merchant fleet for ammonia shipping. At the receiving end, a storage and cracking facility would convert the ammonia into hydrogen and nitrogen with the latter being discharged to atmosphere. The work described in Chapter 6 for the manufacture of ammonia using hydrogen produced by electrolysis indicates that this would be a high-cost option; consequently, this option is developed as a blue hydrogen concept from natural gas with the produced carbon dioxide captured and geo-sequestrated.

[8] Efficiencies over 90% for LNG trains are often quoted but this usually concerns the efficiency of the refrigeration plant and not the whole facility which incorporates liquids separation, a gas plant and storage.

As noted in the earlier chapter, the production cost of ammonia from natural gas priced at $2.GJ (HHV) is about $304/t which implies the hydrogen cost of production of $1,722/t. For production at the remote site being considered, this will increase the capital cost by 50% and since the operating costs are considered as a function of the capital cost, the non-feedstock operating costs will also rise. The estimate for the production of ammonia for this option is $323.5/t or $1,833/t for the hydrogen component.

For 850 kt/y of ammonia (150 kt/y of hydrogen), 1.48 Mt/y of carbon dioxide would be produced, which would be captured and stored underground. If the cost of this is $50/t, this will raise the cost of producing a 'blue' ammonia to $410.6/t. If the shipping cost is $60/t, the landed cost is then $470.6/t or $2,666/t for the hydrogen component.

To this is added the cost of the receiving terminal and the ammonia cracking facility, which adds a further $295/t to the cost giving a total estimate for this option as $2,961/t or $24.8/GJ.

As discussed in the Chapters 6, the use of naphthenes (Spera Process) as a transport agent would be expected to have similar economics.

Summary

In summary the three options considered here are as follows:

- Hydrogen production from natural gas with CCS and shipping as liquid — $44.69/GJ
- LNG shipping and hydrogen production and CCS at receiving end — $36.7/GJ
- Ammonia production with CCS, shipping and cracking — $24.8/GJ

This clearly shows the benefit of using an intermediate to transport hydrogen (ammonia or naphthene). The LNG result is probably conservative as it should be feasible to purchase LNG on a long-term contract lower than the landed price ($10/GJ) used in this analysis.

Case Study 4: Hydrogen from Victorian Lignite to North East Asia

The Latrobe Valley's lignite deposits in Victoria, Australia, is one of the world's largest single point sources of carbon. The coal beds are extensive, very thick (300 m in places) and have minimal overburden. Production costs are very low. Furthermore, they are adjacent to the Bass Strait oil and gas province, which nowadays contains large depleted oil and gas reservoirs which could be used for carbon geo-sequestration.[9] The main issue with this lignite deposit is the very high moisture content. When dried out it becomes pyrophoric which makes it difficult to export without extensive pre-treatment.

There is the option to use this resource for the production of hydrogen and export the hydrogen to Northeast Asia and simultaneously use the depleted gas wells of the Bass Strait to store the co-produced carbon dioxide. Two options are evaluated (i) the production of hydrogen, liquefaction and shipping as liquid[10] and (ii) the production of liquid ammonia and shipping as ammonia for conversion into hydrogen at the destination.

A typical ultimate analysis of Latrobe Valley lignite is given in Table 8.7.

As can be deduced from the high water content shown in this analysis, there will have to be extensive plant for coal handling and drying prior to gasification. Using the data given earlier for a National Academy of Science (NAS) coal study[11] the estimate for a plant using a lignite source is developed in Table 8.8. This shows a 50% higher capital cost for the lignite feedstock compare to a black coal operation.

An allowance is made for an urban Australian location by increasing the capital estimate by a further 30%. The estimate for the project is

---

[9] The CarbonNet Project: https://earthresources.vic.au/projects/carbonnet-project

[10] Hosie E. (22 July 2019) Kawasaki begins construction of coal to hydrogen plant. *Australian Mining.*

[11] The Hydrogen Economy Opportunities, Costs, Barriers and R&D Needs. National Academy of Sciences, 2004

Table 8.7: Ultimate analysis of Latrobe Valley lignite.

Ultimate analysis (Dry Ash Free)		
Carbon	wt.%	66.90%
Hydrogen	wt%	4.80%
Oxygen	wt%	27.50%
Nitrogen	wt%	0.51%
Sulphur	wt%	0.26%
		99.97%
Ash (as received)	wt%	7.40%
Moisture (as received)	wt%	64.70%
As received Basis		
Carbon	wt.%	18.67%
Hydrogen	wt%	1.34%
Oxygen	wt%	7.67%
Nitrogen	wt%	0.14%
Sulphur	wt%	0.07%
Ash (as received)	wt%	7.40%
Moisture (as received)	wt%	64.70%
		99.99%

Table 8.8: Capital cost estimate for a lignite coal gasification process (US Gulf location; 2004$).

Capex	NAS	Lignite
Coal handling	95.00	190.00
Gasifier	173.00	346
Air separation	151.00	151.00
CO SHIFT etc.	208.00	249.6
Claus plant	40.00	40.00
	667.00	976.60
General facilities (30%)	200.10	292.98
Engineering/start up (10%)	66.70	97.66

(*Continued*)

Table 8.8: (*Continued*)

Capex	NAS	Lignite
Contingencies (10%)	66.70	97.66
Land (7%)	46.69	68.36
	1047.19	1533.26

Table 8.9: Estimate of hydrogen production cost from Latrobe Valley lignite.

		NAS	VIC Lignite
Nominal hydrogen production	kt/y	408.00	408.00
	kt/d	1.200	1.200
	PJ/y	48.94	48.94
Construction period	y	4	4
Required return	%	10%	10%
Operating period	y	20	20
Capital recovery	%	15.15%	15.15%
Base Capital Cost (adjusted)	M$	1,047.19	1,533.26
Base Year		2004	2004
Location factor		1	1.3
NF index		1.558	1.558
FINAL CAPEX	M$	1,632.00	3,106.37
Capital Charges	M$/y	247.23	470.58
CAPITAL CHARGES (C)		247.23	470.58
Operating Costs			
Labour (1% Capex)	M$/y	40.80	77.66
Maintenance (3% Capex)	M$/y	57.12	108.72
Insurance (1.5% Capex)	M$/y	24.48	46.60
Catalysts and Chemicals (1% Capex)	M$/y	16.32	31.06
NON-FEED OPEX	M$/y	138.72	264.04
Non Gas Feedstocks			
Electricity	MWh/y	198,300	198,300
Non-Gas Feedstock Costs (power @ $50/MWh)	M$My	9.91	9.91

Table 8.9: (*Continued*)

		NAS	**VIC Lignite**
TOTAL NON FEED COSTS	M$/y	395.86	744.54
COAL USAGE	PJ/y	72.31	150.00
Efficiency	%	67.68%	32.63%
COAL PRICE	$/GJ	1.00	1.00
COAL COSTS	M$/y	72.31	150.00
UNIT PRODUCTION COST CALCULATION			
Gross costs	M$/y	468.17	894.54
By-product Credits	M$/y	5.34	0.00
Net costs	M$/y	462.84	894.54
UNIT HYDROGEN PRODUCTION COST	$/t	1,134.40	2,192.49
	$/GJ	9.46	18.28

detailed in Table 8.9, which also includes the black coal version (NAS) for comparison.

Although a Claus plant is maintained in the capital costs, Latrobe Valley lignite contains almost no sulphur and as a consequence no significant by-product credits arise. It is assumed the product hydrogen is used immediately and working capital is considered as zero. The estimate is for a hydrogen production cost of $2192/t or $18.28/GJ (LHV).

This project would produce 23.5 Mt/y of carbon dioxide. If the carbon disposal cost is $50/t, the cost of hydrogen production increases to $5073/t or $42.3/GJ.

This is a very high price for the production of hydrogen. An alternative might be the production of ammonia. An estimate is based on the production of 1.205 Mt/y ammonia is shown in Table 8.10.

It is assumed the lignite is available at $1/GJ and power can be exported from the facility at $50/MWh. The estimate for the production cost of ammonia is $473/t; note this is considerably higher than the cost estimates for the production of ammonia from natural gas. This ammonia cost is equivalent to a hydrogen cost of $2681/t.

Table 8.10: Production of ammonia from Latrobe Valley lignite.

Nominal ammonia capacity	kt/y	1,205.40
	kt/d	3.55
	PJ/y	27.12
Construction period	y	4
Required return	%	10.00%
Operating period	y	20
Capital recovery	%	15.15%
Base Capital Cost	M$	1,677.54
Base Year		2007
Location factor		1.3
NF index		1.36
FINAL CAPEX	M$	2,958.09
Capital Charges	M$/y	448.12
Nominal Value of ammonia	$/t	200.00
Working Capital	M$	21.27
Return of Working Capital	M$/y	2.13
CAPITAL CHARGES	M$/y	450.24
Operating Costs		
Labour (3% Capex)	M$/y	8.88
Maintenance (3% Capex)	M$/y	27.23
Insurance (1.5% Capex)	M$/y	11.67
Catalysts and Chemicals (1% Capex)	M$/y	23.60
NON-FEED OPEX	M$/y	71.38
Non-Gas Feedstock Costs	M$/y	0.00
TOTAL NON-FEED COSTS	M$/y	521.62
Coal usage	PJ/y	80.00
Efficiency	%	33.90%
Coal price	$/GJ	1.00
Coal costs	M$/y	80.00

Table 8.10: (*Continued*)

PRODUCTION COST CALCULATION		
Gross costs	M$/y	601.62
By-product credits (B)	M$/y	31.25
Net costs	M$/y	570.37
UNIT AMMONIA PRODUCTION COST	$/t	473.18
HYDROGEN PRODUCTION COST	$/t	2,681.4

This option produces 7.61 Mt/y of carbon dioxide. If disposal cost of carbon dioxide is $50/t the production cost of ammonia rises to $789/t.

If ammonia shipping cost to Northeast Asia is added ($60/t) and storage and cracking cost is added (295/t) the cost of hydrogen to a Northeast Asian destination for this case is $5106/t or $42.7/GJ (LHV).

On present statistics, it is difficult to see any of the options using Victorian lignite would be economically viable. Furthermore, at the time of writing the oil province of the Bass Strait is still producing oil and gas. They therefore have a tangible value to the present owners. How the present operations could cope with carbon geo-sequestration from other parties is not yet clear.

APPENDICES

A1: Properties of Hydrogen
A2: Comparison of Hydrogen to Other Fuels
A3: Conversion Factors
A4: Cost Indices
A5: Location Factors
A6: Relation of Delivered Equipment to Final Capital Cost
A7: Methodology for Economic Analysis
A8: Indexed Fuel and Construction Costs
A9: Accuracy of Cost Estimates
A10: Abbreviations

A1. Properties of Hydrogen[1]

Selected properties of Hydrogen at 20C and 1 atm.		
Molecular Weight		2.016
Molar volume (20C)	L/mol	24.06
Density	kg/m^3	0.838
Specific gravity	Air = 1	0.0696
Viscosity	g/cm/sec	8.81E + 0.05
Diffusivity	m^2/h	1.697
Thermal conductivity	W/m.K	0.1825
Expansion ratio	liquid to gas	847
Boiling point	°C	−253
Specific heat (Cp)	J/g.K	14.29
Specific heat (Cv)	J/g.K	10.16
Specific volume	cm/kg	11.39
Diffusion coefficient in air	cm^2/s	0.756
Enthalpy	kJ/kg	4098
Entropy	J/gK	64.44

Selected combustion properties of hydrogen at 20C and 1 atm.		
Flammability limits in air	vol%	4–75
Flammability limits in oxygen	vol%	4–95
Detonability limits in air	vol%	18–59
Detonability limits in oxygen	vol%	15–90
Minimum ignition energy in air	micro J	19

[1]Adapted from Rajeshwar K, McConnell R, Harrison K, Licht S. (2008) Renewable energy and the hydrogen economy. In: Rajeshwar K, McConnell R, Licht S (eds), *Solar Hydrogen Generation*. Springer, New York, NY. doi:10.1007/978-0-387-72810-0_1.

Auto ignition temperature	C	585
Quenching gap in air	cm	0.064
Diffusion coefficient in air	cm^2/s	0.756
Flame velocity	m/s	2.7–3.5
Flame temperature	°C	2045

Other Physical Properties		
Boiling point	K	20.268
Melting point	K	14.01
Triple point temperature	K	13.8
Triple point pressure	kPa	7.2
Critical point temperature	K	33.25
Critical point pressure	MPa	1.297
Critical point density	kg/cm	31.4

A2: Comparison of Hydrogen to Other Fuels

Comparison of hydrogen and other fuels: Lower heating value (LHV) and 1 atm, 25C for gases

		Hydrogen	Methane	Gasoline	Diesel	Methanol
Density	kg/m^3	0.0838	0.71	702	855	799
Energy density	MJ/m^3	10.8	32.6	31,240	36,340	14,500
Energy density	kWh/m^3	3	9.1	8680	10,090	4030
Energy density	kWh/kg	33.3	12.8	12.4	11.8	5
LHV	kJ/g	119.9	50	44.5	42.5	19.5
Higher heating value (HHV)	kJ/g	141.9	55.5	47.5	44.8	22.7

Specific volume and heating values of fuels

	L/t	bbl/t	HHV GJ/t	LHV GJ/t
Acetylene			49.92	48.17
Ammonia			22.50	18.60
AVTUR	1261	7.93	46.40	
n-Butane			49.50	45.73
iso-Butane			49.41	45.61
Butanes	1928	12.13	49.45	45.67
But-1-ene			48.45	45.31
Butenes			48.10	45.00
Carbon			32.80	32.80
Carbon Monoxide			10.09	10.09
Crude oil (35API)	1177	7.40	45.00	
Crude oil (40 API)	1212	7.62		
Diesel	1182	7.43	45.90	43.00
Dimethyl ether (DME)	1493		31.00	28.40
Ethane	2654		51.88	47.49
Ethylene			50.29	47.15
Fuel Oil (LS)	1110	6.98	44.10	
Gasoline	1360	8.56	46.70	42.50
Hydrogen			141.86	119.93
Liquefied petroleum gas (LPG)		12.35	49.87	46.01
Methanol	1272	8.00	22.70	19.50
Methane			55.56	50.07
Natural gas			53.90	
Naphtha	1534	9.00	48.10	
Propane	1998	12.57	50.33	46.36
Propylene			48.95	45.77
Liquefied natural gas (LNG)	2197.80	13.83	54.4	
TAPIS Crude	1231.07	7.75	43.52	

A3: Conversion Factors

Conversion Factors		
Cubic metre	35.315	Cubic feet
m^3@15C	35.383	Ft^3@60F
GJ=	0.9478	MMBTU
$/GJ=	1.055	$/MMBTU
1 kWh	3.6	MJ
lb	0.4536	kg
HP	0.7457	kW
GWh	3.6E-03	PJ
US Gallon	3.78541	Litre
TEMPERATURE		
ABSOLUTE ZERO	C	−273.15
NORMAL	C	15
GAS CONSTANT	R	1.9859
	J/mol/K	8.3145
DAYS PER YEAR		340
HOURS PER YEAR		8160
NAUTICAL MILES	1.852	km

Note: M = Mega (million in SI); MM = million US customary units.

A4: Cost Indices

Developed from Nelson–Farrar Cost Indices and Chemical Engineering Process Cost Index

			INDEX = 0.6 MAT + 0.4 LAB		
Year	Material	Equip.	Labour	Index	NF factor
Weight	0.4	0.0	0.6		
1946	100.0	100.0	100.0	100.0	28.576
1947	122.4	114.2	113.5	117.1	24.411
1948	139.5	122.1	128.0	132.6	21.550
1949	143.6	121.6	137.1	139.7	20.455
1950	149.5	126.2	144.0	146.2	19.546
1951	164.0	145.0	152.5	157.1	18.190
1952	164.3	153.1	163.1	163.6	17.469
1953	172.4	158.8	174.2	173.5	16.472
1954	174.6	160.7	183.3	179.8	15.891
1955	176.1	161.5	189.6	184.2	15.513
1956	190.4	180.5	198.2	195.1	14.648
1957	201.9	192.1	208.6	205.9	13.877
1958	204.1	192.4	220.4	213.9	13.361
1959	207.8	196.1	231.6	222.1	12.867
1960	207.6	200.0	241.9	228.2	12.523
1961	207.7	199.5	249.4	232.7	12.279
1962	205.9	198.8	258.8	237.6	12.025
1963	206.3	201.4	268.4	243.6	11.733
1964	209.6	206.8	280.5	252.1	11.333
1965	212.0	211.6	294.4	261.4	10.930
1966	216.2	220.9	310.9	273.0	10.467
1967	219.7	226.1	331.3	286.7	9.969
1968	224.1	228.8	357.4	304.1	9.397
1969	234.9	239.3	391.8	329.0	8.685
1970	250.5	254.3	441.1	364.9	7.832
1971	265.2	268.7	499.9	406.0	7.038

(*Continued*)

				INDEX = 0.6 MAT + 0.4 LAB	
Year	Material	Equip.	Labour	Index	NF factor
1972	277.8	278.0	545.6	438.5	6.517
1973	292.3	291.4	585.2	468.0	6.105
1974	373.3	361.8	623.6	523.5	5.459
1975	421.0	415.9	678.5	575.5	4.965
1976	445.2	423.8	729.4	615.7	4.641
1977	471.3	438.2	774.1	653.0	4.376
1978	516.7	474.1	824.1	701.1	4.076
1979	573.1	515.4	879.0	756.6	3.777
1980	629.2	578.1	951.9	822.8	3.473
1981	693.2	647.9	1044.2	903.8	3.162
1982	707.6	662.8	1154.2	975.6	2.929
1983	712.4	656.8	1234.8	1025.8	2.786
1984	735.3	665.6	1278.1	1061.0	2.693
1985	739.6	673.4	1297.6	1074.4	2.660
1986	730.0	684.4	1330.0	1090.0	2.622
1987	748.9	703.1	1370.0	1121.6	2.548
1988	802.8	732.5	1405.6	1164.5	2.454
1989	829.2	769.9	1440.4	1195.9	2.389
1990	832.8	797.5	1487.7	1225.7	2.331
1991	832.3	827.5	1533.3	1252.9	2.281
1992	824.6	837.6	1579.2	1277.4	2.237
1993	846.5	842.8	1620.2	1310.7	2.180
1994	877.2	851.1	1664.7	1349.7	2.117
1995	918.0	879.5	1708.1	1392.1	2.053
1996	917.1	903.5	1753.5	1419.0	2.014
1997	923.9	910.5	1799.5	1449.2	1.972

(*Continued*)

(*Continued*)

			INDEX = 0.6 MAT + 0.4 LAB		
Year	Material	Equip.	Labour	Index	NF factor
1998	917.5	933.2	1851.0	1477.6	1.934
1999	883.5	920.3	1906.3	1497.2	1.909
2000	896.1	917.8	1973.7	1542.7	1.852
2001	877.7	939.3	2047.7	1579.7	1.809
2002	899.7	951.3	2137.2	1642.2	1.740
2003	933.8	956.7	2228.1	1710.4	1.671
2004	993.8	1112.7	2314.2	1833.6	1.558
2005	1179.8	1062.1	2411.6	1918.8	1.489
2006	1273.5	1113.3	2497.8	2008.1	1.423
2007	1364.8	1189.3	2601.4	2106.7	1.356
2008	1572	1230.6	2704.3	2251.4	1.269
2009	1324.8	1239.7	2813	2217.7	1.289
2010	1480.1	1224.7	2909.3	2337.6	1.222
2011	1610.5	1256.4	2985.6	2435.6	1.173
2012	1579.7	1286.1	3055.6	2465.2	1.159
2013	1538.7	1317.5	3123.4	2489.5	1.148
2014	1571.8	1335.8	3210.7	2555.2	1.118
2015	1350.3	1434.9	3293.8	2550.2	1.121
2016	1403.1	1340.7	3395.8	2598.7	1.100
2017 JUL	1519.3	1519.3	3468.7	2688.9	1.063
2018				2857.58	1.000

A5: Location Factors[2]

Table A5-1: Capital (CAPEX) and operating cost (OPEX) factors.

	1	2	3	4
Climate/Terrain	**Benign**	**Difficult**	**Difficult**	**Extreme**
Gas Transmission	Present	Present	No	No
Fresh Water	Present	Present	No	No
Ship Loading	Present	Present	No	No
Employee Housing	Present	Present	No	No
Labour Costs	Low	High	High	High
RELATIVE CAPEX	1.000	1.155	1.562	2.250
RELATIVE OPEX	1.000	1.139	1.520	2.039
Examples	US Gulf	Urban Australia	Remote FE	Offshore
	Canada	New Zealand	Remote Aus	Arctic
		Developed FE		
		Middle East		

A6: Relation of Delivered Equipment to Final Capital Cost[3]

Table A6-1: Relation of delivered equipment to final capital cost.

	Solids (%)	Solid-fluid (%)	Fluid (%)
Equipment	100	100	100
Installation	45	39	37
Piping	16	31	93
Structural foundations			7
Electrical	10	10	10
Instruments	9	13	

(Continued)

[2] Assessment of Cost Benefits of Flexible and Alternative Fuel Use in the US Transportation Sector — Technical Report Three — Methanol Production and Transportation Costs, US Department of Energy, November 1989.

[3] Perry RH, Green DW. (1997) *Perry's Chemical Engineers' Handbook*, 7th ed. McGraw-Hill. Table T 9-51.

Table A6-1: *(Continued)*

	Solids (%)	Solid-fluid (%)	Fluid (%)
Building services	25	39	26
Site preparation	13	10	10
Auxiliaries	40	55	70
Sub total plant	258	297	353
Field expense	39	34	41
Engineering	33	32	33
Direct plant costs	330	363	427
Contractor fees	17	18	21
Contingency	34	36	42
Capital Investment	**381**	**417**	**490**

Note that the equipment cost is about one quarter of the final cost. Often when a new process or product comes to market, the initial cost is very high. As time progresses, manufacturers of the product continually improve production techniques and introduce improvements in the efficiency of production. Production efficiencies are also improved by the scale of production — the more gadgets produced lowers the fixed cost of the production. The result is that the cost of the product falls. This can be quite dramatic in the early days of the introduction of a new product. This phenomenon is well known and often referred to as a learning curve. Figure A6-1.

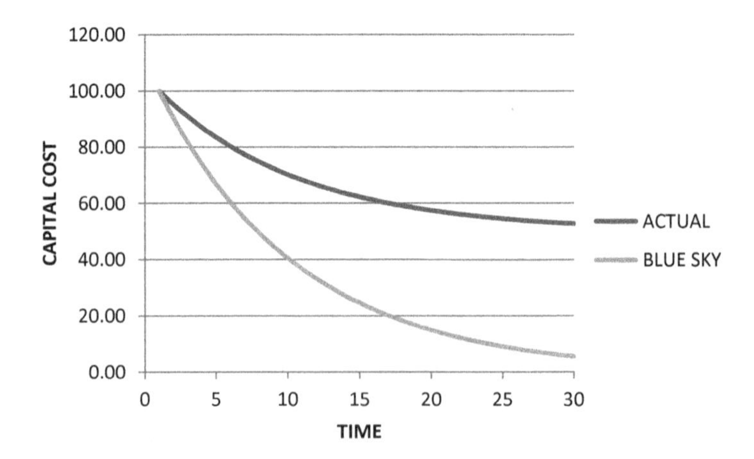

Fig. A6-1: The learning curve.

However, there are limits to the extent of this fall that over time levels off and reaches an asymptote. All other things being equal, the level of the asymptote (lowest production cost) is determined by:

- The cost of labour to assemble the item
- The cost of energy used to produce the item
- The cost of other material in the construction of the item

In writing this book the author has come across several articles that appear to assume a constant fall in the capital cost of a specific item so that at some future date a project becomes profitable. This particularly concerns the cost of renewable power, especially from PV systems, where 10-year projections of cost predict very low power generation cost. It is the view of the author that this is a blue-sky scenario as it assumes the fall in the cost of the solar voltaic cell is replicated in the cost structure of the system, which is only a small part of the finished system.

A7: Methodology for Economic Analysis

What is required is a rapid approach to the determination of the economic viability of a particular technology of interest, namely a concept analysis where speed is not gained at the expense of accuracy. This requires a systematic approach in which various technologies and approaches are treated in the same manner so that the economics from one route to hydrogen can be compared to another.

The methodology described was devised by ICI PLC in order to evaluate all of the diverse routes to the production of ethylene from any feedstock using widely disparate technologies with different plant construction periods and lives of operation. The methodology has been published by Stratton *et al*[4] and is has generally applicable for energy intensive industries. The basic economic equation is

$$\mathbf{P = F + C + O}$$

Where P, the unit production cost of the product of interest (hydrogen say), is equal to the sum of the unit feedstock costs (F), the unit capital

[4] Stratton A. (1982) *A Simplified Method of Calculating Product Cost.* Technical Note 3, Economic Assessment Service, IEA Coal Research, London.

costs (C) and the unit non-feedstock operating costs (O). This can be expressed as a fixed-variable equation with the fixed part of the equation representing the return on capital (ROC) (the unit capital costs, C, independent of tax considerations) together with all the unit non-feedstock operating costs (O).

Capital Costs (C)[5]

The capital costs are developed for greenfield projects completely isolated from other facilities. All the costs associated with utilities (unless otherwise accounted) are allowed for in the capital cost. Some processes require small amounts of power. This is considered as an import.

Capital is estimated using published information and using the location factors and Nelson–Farrer Indices given above is adjusted to US Gulf site and 2018 costs for all processes. Scaling used the exponent method namely:

Capital of Plant [1]/Capital of Plant [2] = {Capacity of Plant [1]/ Capacity of Plant [2]}^{n}

Where n is a constant with a typical value, which is typically 0.7.

Capital cost of facilities in other locations can be estimated from the factors given in Appendix A5.

Capital Recovery Factors

For a plant with a capital cost of Co, the plant investment cost, C, capitalises the return on investment during construction of the plant, that is takes account of expenditure and the required return during the construction period.

$$C = \sum_{s=0}^{p} a_s (1+i)^{p-s}$$

[5] Brown TR, Capital cost estimating, Hydrocarbon Processing, pp. 93–100, October 2000. and Summerfeld JT, Petrochemical plant costs for the new millennium, Hydrocarbon Processing 80(6):103–108, June 2001.

Where a_s represents the breakdown of capital expended over the construction period. p is the first year of production, and s is a general year of the project starting s = 0, with construction complete at s = p. The return on investment is i. The values of a_s are given in Table A7-1:

Table A7-1: Values for a(s).

Construction Period	**1 year**	**2 years**	**3 years**	**4 years**	**5 years**
Values of a(s)					
a(0)	100%	50%	30%	17%	4%
a(1)		50%	45%	32%	14%
a(2)			25%	26%	32%
a(3)				25%	36%
a(4)					14%

The general Discounted Cash Flow (DCF) equation can be written:

$$C = \sum_{r=1}^{N} (Rr - FCr - VCr)/(1+i)^r$$

where r is the production year, with N the final production year and Rr is the total product revenue in year r, FCr is the fixed costs in year r, VCr is the variable cost in year r.

This equation is simplified by assuming that there is no build up to full production and full production is achieved as soon as construction is complete. This is followed by N years of full production. Hence:

$$C\,(1+i) = (Rr - FCr - VCr)\sum_{r=1}^{N} 1/(1+i)^r$$

This is rearranged to give:

$$\mathbf{Rr = FCr + VCr + K(1+i)\,C}$$

where K is the sum of the geometric series:

$$\mathbf{K = i\,(1+i)N/[(1+i)N - 1]}$$

Values of K for various values of i (required rate of return) and N (operating life) are given in Table A7-2:

Table A7-2: Values for K.

N	10	15	20	25	30
Interest (i)					
5.00%	0.1295	0.0963	0.0802	0.071	0.0651
7.50%	0.1457	0.1133	0.0981	0.0897	0.0847
10.00%	0.1627	0.1315	0.1175	0.1102	0.1061
12.50%	0.1806	0.1508	0.1381	0.1319	0.1288

The capital recovery factor (K_o) is then:

$$\mathbf{K_o = K\,(1 + i)\,C}$$

and from the table we get:

$$K_o = k(1+i)\,Co \sum_{s=0}^{p} a_s(1+i)^{p-s}$$

Values for the ROC or Ko/Co are given in Tables A7-3 and A7-4 for a royalty free basis and one encompassing a 2% royalty to the process licensor respectively. Table A6-1 has been used for typical non-process items (pipelines, ships etc.) and Table A7-4 for licensed processes.

Table A7-3: Values for return on capital (royalty free basis).

1-year construction						
	N	10	15	20	25	30
	Interest(i)					
	5.00%	13.60%	10.12%	8.43%	7.45%	6.83%
	7.50%	15.66%	12.18%	10.54%	9.64%	9.10%
	10.00%	17.90%	14.46%	12.92%	12.12%	11.67%
	12.50%	20.32%	16.96%	15.54%	14.84%	14.49%
	15.00%	22.91%	19.67%	18.37%	17.79%	17.51%
2-year construction						
	N	10	15	20	25	30
	Interest (i)					
	5.00%	13.94%	10.37%	8.64%	7.64%	7.00%

Table A7-3: *(Continued)*

	7.50%	16.25%	12.64%	10.94%	10.01%	9.44%
	10.00%	18.80%	**15.19%**	**13.57%**	12.72%	12.25%
	12.50%	21.59%	18.02%	16.51%	15.77%	15.39%
	15.00%	24.63%	21.14%	19.75%	19.12%	18.83%
3-year construction						
	N	10	15	20	25	30
	interest (i)					
	5.00%	14.32%	10.65%	8.87%	7.85%	7.19%
	7.50%	16.92%	13.16%	11.39%	10.42%	9.83%
	10.00%	19.84%	**16.02%**	**14.32%**	13.43%	12.93%
	12.50%	23.08%	19.27%	17.65%	16.86%	16.45%
	15.00%	26.68%	22.90%	21.39%	20.71%	20.39%
4-year construction						
	N	10	15	20	25	30
	Interest (i)					
	5.00%	14.59%	10.85%	9.04%	7.99%	7.33%
	7.50%	17.39%	13.52%	11.71%	10.71%	10.11%
	10.00%	20.58%	16.62%	14.85%	13.93%	13.41%
	12.50%	24.17%	20.18%	18.48%	17.66%	17.23%
	15.00%	28.20%	24.21%	22.61%	21.90%	21.56%
5-year construction						
	N	10	15	20	25	30
	Interest (i)					
	5.00%	14.71%	10.94%	9.11%	8.06%	7.39%
	7.50%	17.61%	13.69%	11.85%	10.84%	10.23%
	10.00%	20.91%	16.89%	15.09%	14.16%	13.63%
	12.50%	24.66%	20.58%	18.85%	18.01%	17.58%
	15.00%	28.87%	24.78%	23.15%	22.42%	22.07%

Table A7-4: Values for return on capital with 2% royalty

	N	10	15	20	25	30
1-year construction						
	Interest (i)					
	5.00%	13.87%	10.32%	8.59%	7.60%	6.97%
	7.50%	15.97%	12.42%	10.76%	9.84%	9.28%
	10.00%	18.26%	14.75%	13.18%	12.36%	11.90%
	12.50%	20.73%	17.30%	15.85%	15.14%	14.78%
	15.00%	23.37%	20.06%	18.74%	18.15%	17.86%
2-year construction						
	N	10	15	20	25	30
	Interest (i)					
	5.00%	14.22%	10.58%	8.81%	7.79%	7.14%
	7.50%	16.57%	12.89%	11.16%	10.21%	9.63%
	10.00%	19.17%	**15.49%**	**13.84%**	12.98%	12.50%
	12.50%	22.02%	18.38%	16.84%	16.09%	15.70%
	15.00%	25.13%	21.56%	20.15%	19.51%	19.20%
3-year construction						
	N	10	15	20	25	30
	Interest (i)					
	5.00%	14.61%	10.87%	9.05%	8.00%	7.34%
	7.50%	17.26%	13.42%	11.62%	10.63%	10.03%
	10.00%	20.23%	**16.34%**	**14.60%**	13.70%	13.19%
	12.50%	23.54%	19.65%	18.00%	17.20%	16.78%
	15.00%	27.21%	23.36%	21.82%	21.13%	20.80%
4-year construction						
	N	10	15	20	25	30
	Interest (i)					
	5.00%	14.88%	11.07%	9.22%	8.15%	7.47%
	7.50%	17.74%	13.79%	11.94%	10.92%	10.31%

Table A7-4: (*Continued*)

	10.00%	20.99%	16.96%	15.15%	14.21%	13.68%
	12.50%	24.66%	20.58%	18.85%	18.01%	17.58%
	15.00%	28.77%	24.69%	23.06%	22.33%	21.99%
5-year construction						
	N	10	15	20	25	30
	Interest (i)					
	5.00%	15.00%	11.16%	9.29%	8.22%	7.53%
	7.50%	17.96%	13.96%	12.09%	11.06%	10.44%
	10.00%	21.33%	17.23%	15.39%	14.44%	13.90%
	12.50%	25.15%	20.99%	19.23%	18.37%	17.93%
	15.00%	29.45%	25.28%	23.61%	22.87%	22.51%

The selection of a rate of capital return is dependent on many factors including the nature of the industry in question. For upstream oil and gas developments, or relatively small-scale process plant, high rates of capital return are often demanded by the investors to offset short operational lives or perceived higher levels of risk. For very long-term (30 year) infrastructure projects often accessing government funds, far lower rates of return are required. Many Greenfield operations in the process industries are planned for a lifetime of 15–20 years and rates of return are as appropriate. Commonly used values for the ROC in this work are emboldened in the tables.

Fixed Operating Costs (O)

Working capital

Rather than capitalise the working capital and handling it with the project capital (Stratton), the working capital is treated as an annual operating cost. The reasoning behind this is that working capital is normally borrowed against the business and is fully recovered at the end of the project. The outgoings are the interest on the debt. The value of working capital can be taken as 5% of the plant capital or 30 days stock. The latter is generally smaller than the former and was used when sufficient data permitted its calculation.

Labour, maintenance and administrative costs

As a general rule, labour and maintenance were each charged at the rate 3% of the capital per annum. For labour, this included both direct and indirect labour costs. For maintenance this included both materials and labour. Over the past decades many companies have made attempts to reduce the operating labour and maintenance charges. Labour can be reduced by extensive computer control. However, the success or otherwise, in reducing the maintenance charge is difficult to quantify, several operations have suffered major problems claimed to be due to the cutbacks in maintenance costs. Administrative costs are basically insurance and local land taxes. A value of 1.5% of the fixed capital as an annual charge was used.

Catalysts and chemicals

Most plants require some chemicals for water treatment purposes. Catalyst charges are based on a 3–5-year turnaround.

Other operating costs

For some processes require inputs other than the principal hydrocarbon feed. This is usually electric power and typical average values were used.

A8: Indexed Fuel and Construction Costs

For the most part we are concerned with hydrocarbon feedstock that is related to the prevailing crude oil price. For some feedstock and hydrocarbon by-product this is a strong linear relationship.

We are also concerned with how construction costs change with time. Work by Parker HW[6] has shown relationship in refinery operations

[6] Parker HW, (4 August 2008) Cost shift signals changes in energy investment, use *Oil & Gas Journal.*

between the construction cost index and the refinery fuel cost index. Plotted in the logarithmic form, this relationship has a high linearity with a slope of approximately unity. The relationship is illustrated in Figure A8-1.

Since we know that most fuels show a linear correlation with the prevailing crude oil price, we can develop a construction cost crude oil price relationship that is illustrated in Figure A8-2. This shows a correlation plot of the construction cost index against an index based on the price of WTI

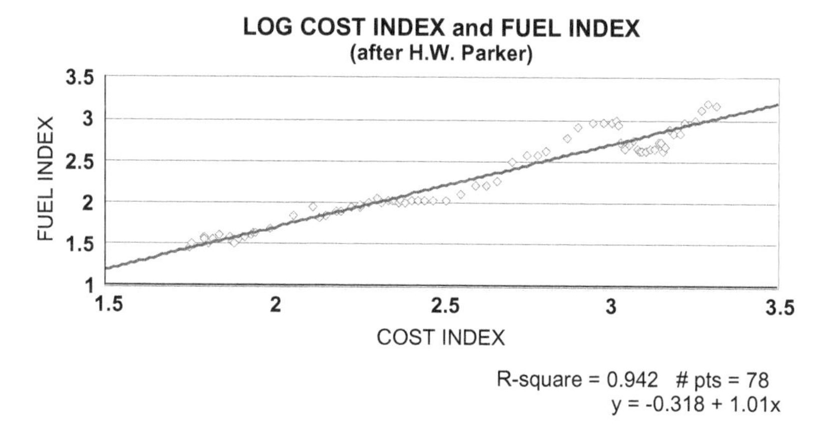

Fig. A8-1: Plot of fuel index against cost index.

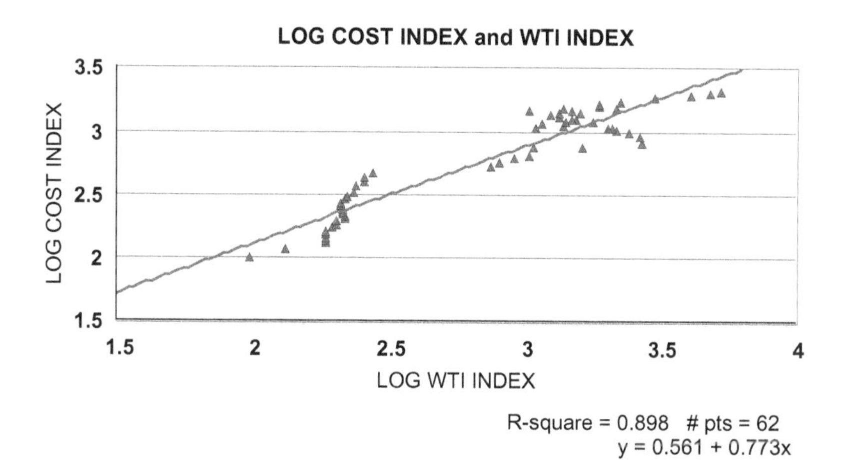

Fig: A8-2: Plot of Cost index against WTI index.

crude oil. As may be expected, there is more variation in this correlation but it still shows a correlation factor or nearly 0.9.

Using this correlation, we can impute crude oil price corresponding to a particular construction cost index. Using the 2007 value for the construction cost implies an equivalent oil price of $70 per barrel. This is used as the base price or oil and derivatives in the analysis.

A9: Accuracy of Cost Estimates

Cost estimates fall into four types: Concept, Feasibility, Bankable Feasibility and Front End-Engineering and Design. These types of study are performed at different stages of a project as illustrated in Figure A9-1.

This book concerns the first concept stage from which an optimum concept can be defined and the key issues for project feasibility identified.

Concept or Scoping Study

OBJECTIVE: To generally define the project

- Define the product and by-products
- Identify feedstock and price range

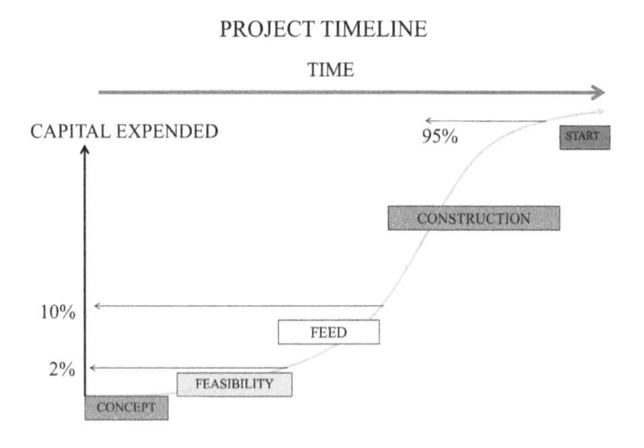

Fig. A9-1: Project timeline and different levels of cost estimation.

- The product market and sales price range
- The project scale
- The technology to be employed
- Provisional mass and energy balance for process
- Identify process licensors/ EPC contractors
- Potential location
- Order estimate for CAPEX and OPEX
- Viability of the concept

STUDY LOCATION: Home office;

COST: <$100,000;

ERROR in cost estimate: +/– 25%–40%

Feasibility Study

OBJECTIVE: To define a specific project

- Define the product and by-products in terms of purity etc.
- Define the product market and outline off-take agreements
- Identify feedstock source, quality and supplier and obtain outline price agreements
- **E**ngineering, **P**rocurement and **C**onstruction contractor appointed
- The project location and scale is defined
- The technology to be employed identified and supplier details are obtained
- Mass and energy balance of all major unit operations are defined
- Budget cost estimates for major items obtained
- Major units are sized, constructors identified and installation defined
- Utilities and off-sites are defined and costed
- Estimate for CAPEX and OPEX

STUDY LOCATION: Home office and EPC contractor office

COST: 1%–2% of final CAPEX

ERROR in cost estimate: +/–15%

Bankable Feasibility Study

OBJECTIVE: To help in raising capital from third parties
All of the actions for a Feasibility Study plus:

- Written and executed feedstock off-take agreements
- Written and executed product and by-product off-take agreements
- Environmental Impact Statement — approvals commence
- Review of Feasibility Study by independent EPC contractor
- Review of project feasibility by independent auditor
- Production of *pro-forma* revenue and balance sheet
- Production of *Product Disclosure Statement*
- Preliminary discussions with potential investors

STUDY LOCATION: Home office/Investment advisor office

COST: to 3 or more % final CAPEX; Can be high for large projects.

ERROR in cost estimate: +/−15% !

Front-End Engineering and Design

OBJECTIVE: To Finalise CAPEX and OPEX prior to Final Investment Decision

- All Government Approvals
- EPC contractor produces engineering design for all major items of equipment
- EPC contractor identifies manufactures of equipment
- EPC contractor obtains cost estimates for equipment procurement
- EPC contractor defines and engineers major utilities and off-sites
- EPC contractor defines construction logistics
- EPC contractor produces final estimate for CAPEX and OPEX

STUDY LOCATION: EPC contractor home office

COST: 10% or more of final CAPEX; Can be higher for large projects

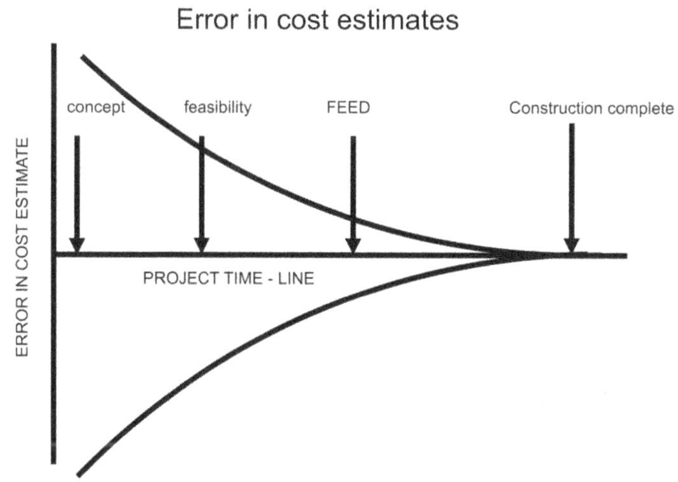

Fig. A9-2: Plot of Error in cost estimate against project timeline

ERROR in cost estimate: +/–5%

THE ERROR IN THE COST ESTIMATE IS INVERSELY PROPORTIONAL TO THE MONEY SPENT: Figure A9-2.

A10: Abbreviations

AVTUR	Aviation turbine fuel — Jet fuel
a	Annum (year)
A	Amp
AC	Alternating Current
AFC	Alkaline Fuel Cell
bbl	Petroleum barrel
bbl/d	Barrels per day
bcfd	Billions of cubic feet per day
BTU	British Thermal Unit
BTX	Benzene, Toluene and Xylene mixture (can include ethylbenzene; BETX)

C	Coulomb
°C	Degrees Centigrade (Celsius)
Capex	Capital cost
cf	Cubic foot
CIF	Container, insurance and freight (destination port price)
cm	Cubic metre
CGS	Carbon Geo-sequestration
CNG	Compressed Natural Gas
DC	Direct Current
DME	Dimethyl ether
EOR	Enhanced Oil Recovery
°F	Degrees Fahrenheit
FOB	Free on board (embarkation port price)
FSRO	Floating Storage and Regasification Unit
GHG	Greenhouse Gases
GJ	Gigajoule (10E+9 joules)
HE	Heat Exchanger
HHV	Higher heating value (gross)
HP	Horse power
HPT	High Pressure Turbine
HTEC	High Temperature Electrolysis Cell
HTWGS	High Temperature Water-Gas-Shift
J-T	Joule Thompson (valve, effect)
K	Degrees Kelvin (absolute temperature scale)
kt/y	Thousand metric tonnes per year
kW	Kilowatt
kWh	Kilowatt hour
L	Litre
lb	Pound

LDPE	Low-density polyethylene
LHV	Lower heating value (net)
LLDPE	Linear low-density polyethylene
LNG	Liquefied natural gas (mainly methane)
LPG	Liquefied petroleum gas (usually propane and butane)
LPT	Low-Pressure Turbine
LTWGS	Low-Temperature Water-Gas-Shift
MCH	Methyl-*cyclo*-Hexane
MEA	Methyl ethanolamine
MM$	Million US dollars (2007)
MMBTU	Million (US Customary) BTU
MN	Methane Number
Mt	Million metric tonnes
MWS	Municipal Waste Stream
NGL	Natural gas liquids (ethane, propane, butane)
Opex	Non-feedstock operating costs
PAFC	Phosphoric Acid Fuel Cell
PEM	Proton Exchange Membrane
PJ	Peta joule (10E + 15 joules)
PONA	Paraffins, olefins, naphthenes and aromatics
PSA	Pressure Swing Adsorption
PWR	Pressurised Water Reactor
ROC	Return on Capital
S.I.	Système International d'Unités; metric units
SMR	Steam Methane Reformer
SOE	Solid Oxide Electrolysis
t	Metric tonne
TSA	Temperature Swing Adsorption
t/y	Metric tonnes per year

V	Volt
VGO	Vacuum gas oil
W	Watt
WGS	Water-Gas-Shift
WI	Wobbe Index

INDEX

CPSIA information can be obtained
at www.ICGtesting.com
Printed in the USA
JSHW040405020622
26223JS00001B/28